Richard Henebry

A Contribution to the Phonology of Desi-Irish

An Introduction to the Metrical System of Munster Poetry

Richard Henebry

A Contribution to the Phonology of Desi-Irish
An Introduction to the Metrical System of Munster Poetry

ISBN/EAN: 9783744734523

Printed in Europe, USA, Canada, Australia, Japan

Cover: Foto ©Thomas Meinert / pixelio.de

More available books at **www.hansebooks.com**

A CONTRIBUTION

TO THE

PHONOLOGY OF DESI-IRISH

TO SERVE AS

AN INTRODUCTION TO THE METRICAL SYSTEM

OF

MUNSTER POETRY.

DISSERTATIO INAUGURALIS

QUAM

AMPLISSIMI PHILOSOPHORUM ORDINIS

CONSENSU ET AUCTORITATE

AD

SUMMOS IN PHILOSOPHIA HONORES

IN

UNIVERSITATE GRYPHISWALDENSI

DIE III. MENSIS AUGUSTI ANNO MDCCCXCVIII

HORA XI

RITE CAPESSENDOS

SCRIPSIT

RICARDUS HENEBRY

LISMORENSIS.

GRYPHISWALDIAE

TYPIS IULII ABEL,

MDCCCXCVIII.

Dom aiti múinte,

don Doctúir Heinrich fial Zimmer,

.|. cenn dinaib prim-ecnaidib fri senchus

isindaimsir seo,

do charait donaib clannaib Goidel

ocus

cara damsa

Abbreviations.

1. The following speakers are quoted:—
   My Father, *Piarus de Henebre*, Killown and Mount Bolton.
   Ph. H., *Pilib de Henebre*, my Uncle, Killown and Mount Bolton.
   M. song, some folk-song of my Mother's, *Eblín ní Chassín, Cloichín an mhargadh*, in South-west Tipperary and Mount Bolton.
   Seaán de Bháll, Mount Bolton.
   S. R., *Seaán Ruadh o'Siodhchán*, Colligan and Mount Bolton.
   T. L., *Tomás o Leannán*, Stradbally and Portlaw.
   P. Walsh, Coolfin, (Portlaw).
   P. Crowley, Portlaw.
   Bob, Robert Weldon, Comeragh.
   Com. song, some folk-song reported by me in Comeragh.
   Thomas Kermode, Bradda near Port-Erin, Isle of Man.
2. Mss.
   O'Daly Ms., a good paper Ms. (1830?) given me by Father Carrigan of Ossory. It was once in the possession of O'Daly who edited the Munster Poets series.
   Ren., the Renehan collection of Mss. in Maynooth College.
   Mur., the very large Bishop Murphy collection in the same College.
   O'Neil Mss. written by Patrick O'Neil, Owning, Piltown, Co. Kilkenny, about 1820.
   S. na Sróna, Sémus na Sróna de Poer, first cousin to *Caitlín de Poer*, Countess of Tyrone (1750).
   M. ní Dhonagán, a celebrated *bean-chaointe* of the Dēsi in the last century. Quoted by O'Neil.
   Dunne collected some folk-songs in Garryricken near Carrick in 1850. O'Neil Mss.
   Brett apud Dunne, someone from whom he reported songs.
   Sheehy, a Lament for the death of Edmund Sheehy who was hanged and beheaded in *Cloichín an mhargadh* in 1776.

Anna, a *caoin* for *Risteird de Poer* of *Garrán an mhuiris,* last century.
Denn Ms., a Ms. of P. Denn of Cappoquin in Waterford College, 1820.

3. Printed works.

Z², Zeuss's Grammatica Celtica. Ed. Ebel.

W., Windisch's Wörterbuch.

P. P., The Poets and Poetry of Munster. Ed. C. P. Meehan C. C. Dublin.

Cass., Statuta Synodalia pro unitis dioecesibus Cassel. et Imelac. (with prayers and instructions in half-phonetic script), Dublin 1813.

Cat., *Teagasg Christuy agus ornihe na mainne agus Trathnona,* Christian Doctrine with morning and evening prayers. Dillon, Cork, circa 1820.

T. G., The Pious Miscellany by *Tadg Gaolach,* Dublin 1868.

Denn, Appendix to the Pious Miscellany by P. Denn, Cappoquin, Cork, 1841.

C. M. O., *Cúirt an mheadhon-oidche,* 2nd ed. O'Brien, Dublin.

D. R. *Eachtra Ghiolla an immfhorráin* by *Donnchadh Ruadh Mac Conmara.* Dublin 1897.

4. Contractions. ppp., participium perfecti passivi, Thom., Thomond, Des., Desmond, : rhymes with, O. I., old Irish.

Introduction.

In the sixteenth century we are suddenly confronted by a system of Irish prosody based on motives entirely different from those which regulated the verse-building of the traditional school. From that time onward there was concurrent use of the old and new systems, the new ever growing, the old ever waning, until the first half of the present century. The new system finally gained complete mastery when *Sémus mac Cuirtin* of Thomond ceased to write some time after 1847. The following specimen of his style *ay cur slán le gaedhilge* (Ren. 70, Irish Battles after *Cath Gabhradh*), is probably the last example of legitimate Irish versification: —

> *Mór an beud mar chuadh ar fághan ('|' fán)*
> *Ar ndaimh léighionda gan meilleán*
> *'Sa chaint cheolmhilis ba áille blas*
> *Ba fada bhí fa dheighmheas.*

Vid. ib. a dirge of his in the same metre on the death of O'Connell *d'eacc in Genoa san mbliadhain* 1847.

Dr. Keatynge a priest of the Diocese of Lismore who died at an advanced age before 1644[1] used both of these metrical systems. His address to *Tadg ó Cobhthaigh crutaire* begins: (Mur. Vol. II p. 115.)

> *Cia an t-saoi le seinntar an chruit*
> *Le niocthar nimh gach nuadhluit*
> *Tré ghoradh guithbhinn a chláir*
> *Mar shruith bhinn fhoghar orgáin.*

---

[1] For an account of all that can now be gleaned of Keatynge's life see Rev. W. P. Burke on the journal of the Waterford Historical and Archeological Society for April 1895, Vol. I, no, 4.

This composition has the Munster *ceangal* appendage written according to the canons of the new system. He also wrote a poem beginning: —

Om sgeol ar árd mhagh Fáil ní chodlaim oidhche
Is do bhreoig go bráth me dála a pobuil dilis
Giodh rófhada atáid na bhfál re brosgar biodhbhad
Fá dheoig gur fhás a lán don chogal triotha.

O'Daly Ms. p. 15. This one of the earliest exemples of the new prosody presents to our view a highly elaborated metric system. It cannot therefore be regarded as new but rather the result of a long period of development which now almost for the first time makes it appearance in the literature. As folk-poetry it, or something akin to it, must have existed from the earliest times but was excluded from Mss. by the conservatism of the scholastic caste, the members whereof alone possessed the power of writing and failed to report it. The classic prosody certainly an inheritance from the period of the tone-accent barely escaped being submerged through the influence of the later stress accent. This, strong enough to effect profound change in the whole structure of the language, made its presence felt also in poetry. Hence the earliest examples of Irish verse we possess are more or less dependent on stress accent. Cf. Ultan's Hymn W. p. 25.

Ron soéra Brígit | sech drúngu démna —
in chróeb co m-bláthuib | in máthair I'su —
intlácht uaslígaib, | ind rígan rígda —
dia ráth ron broena, | ron soéra Brígit.

From a marginal distich in Sg. Z² 953:

Isácher ingáith innócht | fu fúasna fáirrge findfholt
ib. Z² 954:

huas mo lébrán indlínech | fomchain trírech inna nén.
In later examples, only the second half of the long-line was subject to stress-accent. W. 11.

Genair Patraicc in Nemthur | is éd atfét hi scélaib.

Still later even this exception submitted to the technical rules of the poetic sept. Vid. Hymn edited by Stokes in *Zeit. f. Celt. Phil.* Vol. I. p. 62, where accent asserts itself pretty regularly in the end of the line and compare with the late. Contention of the Bards where it is completely lost.

To suppose therefore that the late prosody borrowed its accentuation from a Germanic source is a mistake. In all periods of the literature we find embedded certain barbed fragments of folk-wit that had pierced the exclusiveness of the schools and stand a running testimony to the continued existence of a simple style of versification which the common people might compose and sing. Cf. Wb. 27[a] *teora tonna torunni*, the legal aphorisms in the Brehon Code, and such wise sayings as *atcota sochell saidhbhre, atcota cuirm carna*, Mur 48 *Roscadha Flain Fina mic ossa Rig Saxon*, p. 351. Also bringing together the two extremes of Irish poetry Ultan's Hymn for instance and the so-called Midnight Court written in 1780 we can discern no essential difference of structure. Wherefore I re-affirm an opinion expressed by me some time ago *Zeit. f. Celt. Phil.* Vol. I, p. 141, that" there was reason to suspect that this simple and effective skeleton of verse-building lived from old times in populur song side by side with the elaborated prosody to reappear afterwards when the folk language became the only literature."

Taking the last example cited from Dr. Keatynge and marking accented vowels in colour we find the following scheme.

. ō . å . å . o . ī .
. . ō . å . å . o . ī .
. ō . å . å . o . I .
. ō . å . å . o . ī .

Here then the verse was woven on a fixed net-work of assonance, there is a regular accent beat and alliteration is abundant. A system so dependent on pronunciation demands for its reading some notions of the phonology of modern

Munster Irish. It was called „Munster Poetry" perhaps from the collection bearing the title „Poets and Poetry of Munster" printed by O'Daly. At any rate though prevalent all over Ireland the system attained its highest degree of perfection in the hands of the 17th and 18th century professional rhymers of that province.

A triple subdivision of Munster is made necessary by the fact that short accented vowels in heavy syllables assume at least three distinct grades of intonation. The members of this partition shall be called 1· Thomond (Clare, Limerick), 2· Desmond (Cork, Kerry), and 3· the country of the Dēsi comprising Waterford and the southern portion of Tipperary. As the last-mentioned division is my native language-territory I shall treat of its dialect specially giving such references to the variant usages in the others as will be helpful towards the reading of the whole body of Munster Poetry.

The district here called the country of the Dēsi or shortly Dēsi, is that comprised by the present Dioceses of Waterford and Lismore. To this add the baronies of Iverk and Ida in Kilkenny. Boundaries: From the Meeting of the three Waters below Waterford to Dunmore, along the coast to Youghal, by the Blackwater to Lismore, over *Cnoc Mael domnaig* to the meeting point of Cork, Limerick and Tipperary at Kilbehenny, along the Galtees to *Sliabh na mban*, along the Walsh Mountains to Tory Hill and to The Meeting of the Three Waters. The inhabitants may be distinguished by their surnames into various races: Irish, Cymric, Danish, Norman and English. In East Waterford the Norman element prevails hence the name *Duthaigh Paorach*, the territory of the Le Poers.

Phonetic Symbols.

§ 1. Small uncials and text will be used to connote broad and slender sounds respectively. This distinction will extend to the vowels also in so far as they suffer variation of timbre. U E I therefore shall be used to denote the broad sound of those vowels. Special reductions of L R are represented by l' r'; zh signifies the weakening of r, a Leinater development which extended South as far as the Suir. C C stand for the spirants arising from K k; Y y, those produced by G D and g d respectively. N is the nasal guttural. A O are not used as those vowels are broad for their whole period of duration. rh = unvoiced r.

	Stops		Spirants		Liquids		
	Silent	voiced	silent	voiced	voiced		
labials	P p	B b	F f	W,V v	M m		
dentals	T	D	S		N	L	
alveolar	t	d	s	zh	n	l	R r r'
palatals	k	g	c	y	ŋ		
gutturals	K	G	C	Y	N		

Vowels, a ā á å, E ē e ə, I ı ī i, U u ū, ḷ m̩ ṇ ṛ, L' M' N' R'.
Diphthongs, iə, uə.

Slurred Diphthongs, au ou î ai Ei.

P B F W M are bilabial sounds. (W in auslaut from bh mh, broad is written v.) Their corresponding slender sounds are either bilabial or labiodental.

D T N L require for their production that the tongue be pressed against the upper teeth and hard gum. For L a lighter pressure is required, the tongue is apparently spread wide and the throat organs are held in the position of a u vowel. It is best reproduced by sounding such a vowel before it the tongue being drawn forward to touch the upper teeth.

t d are produced by the tip of the tongue on the hard gum near the roots of the upper teeth. Contact is broken gradually and an incipient spirans makes itself heard after the consonant.

s = Eng. *sh*, Germ. *sch*,

zh = French *g* in such words as *rouge*,

k like in Eng. *kin*, Ger. *kind*.

g like Eng. *give*, Ger. *gib*. As in the case of t d so k g break contact slowly and produce a slight following spirans.

ŋ = Eng. *winged*, Germ. *ging* but without the auslaut palatal k, g, which is often heard both in English and German.

c = Ger. *ich*.

y = Eng. anlaut *y*, Germ. *j*.

ᴋ like guttural *c* in Eng. *cow*, Ger. *kuh*.

c = guttural *ch* in Ger. *ach*, *kuchen*.

ᴙ = Ger. *g* in words like *wagen*.

ɴ = *ng* in Eng. *longed*, Ger. *lang* but without a following guttural consonant, as sometimes in Eng. *long—g*, Ger. *lang—k*.

l′ is the reduction of l the ordinary slender consonant. It is a *ly* or *yl* sound like Italian *gl* heard in conjunction with y as in *ghleo*. The reduction of ʟ will be distinguished occasionally by l′.

ʀ represents *rr*, and *r* unaffected in anlaut,

r′ its reduction, and slender *r* in auslaut — *aire*, anlaut *gr-*, *pr-*, *br-*.

r the ordinary sound.

Vowels.

a=short open *a* as in Eng. *cat* as pronounced in Ireland, Ger. *kann*.

ā the same sound lengthened. Ger. *malen*.

à = Eng. *what*.

å = Eng. *fall*.

ᴇ, always long is like Ger. ö sounded deep in the relaxed

guttural chamber. It may be attempted by pronouncing $g\bar{e}$ with the g of Eng. *go*.

e = Eng. *bed*, Ger. *bet*.

ē = Eng. *a* in *fate*, *late*, Ger. *lesen*.

ǝ the irrational or colourless vowel = the short vowel after the accent in English words e. g. *evident, wicked,* pronounced naturally not affectedly as when one says *evy-dent, wickeed.*

I ɪ like the short and long vowels in Eng. *quill, queen* respectively, pronounced without the *w* and without lip-rounding thus *k(w)ill k(w)īn.* ɪ I and ɛ are the broad sounds corresponding to i ī and ē slender. I = *i* umlaut of *o; cuir* kɪr′, from *cor.* I when not = ɪ + y represents the *i* umlaut of an old diphthong now written *aoi.* ɛ represents the sound of the digraph *ao.*

u = Eng. *puss*, Ger. *muss.*

ū = *oo* in Eng. *cool.*

u stands for ū after y. iú = *ew* in English words like *few, new.*

ḷ. m̥, n̥, r̥, are liquid sonants like *l n* in Eng. *buckle token,* etc. i. e *bakl′, tōkn.*

au, eu, î, ai, ɛi, are long slurred diphthongs or coalescences which can be apparently resolved into the simple sounds which constitute their several signs. î may be imitated by slurring öi. Those arise from short vowels under certain conditions of accent and consonant accompaniment. Vid. § 4, sqq.

au like Ger. *haus,* more open than the *English* pronunciation of *ou* in *house.* It is always strongly nasalized.

ou = the *Irish* pronunciation of *ou* in *house* or of *o* in *bold* i. e. *bould.*

ai like the *English* pronunciation of the pronoun *I*

i = öi like the *Irish* pronunciation of the pronoun *I*

ɛi like the preceding sound but with an *e* instead of an *i* colour.

The numerous digraphs asising from the *caol-leathan* rule will be represented by their simple values.

Of Accent.

§ 2,1. In the prehistoric period a revolution of accent took place in Irish which compared with the original Indogermanic accent shows two remarkable characteristics. One has reference to the place the other to the quality of the accent. With regard to the place we find that the free Indogermanic accent which could rest either on the root or on the ending, has become fixed in the great majority of cases in O. I. Its place in nouns, simple and compound, and simple verbs, was on the first syllable, in the case of compound verbs the position charged from the first to the second element according to certain established laws. As regards quality the Irish accent consisted of stress whereas the Indogermanic represented a chromatic tone change. This stress, being of a very strong expiratory character gave rise to certain phenomena which may be briefly summarized thus: —

The toned syllable retained its inherited vocalic quality and quantity whereas unaccented syllables suffered a qualitative and quantitive weakening. The syllable after the accent became weakest and if not final usually lost its vowel and contracted. *maidin* gen. *maidne, tabhair*, but *taibhrem = taibherem*. Vid. Zimmer, Ueber altirische Betonung, Berlin 1884, and Thurneysen, L'accentuation de l'ancien verbe irlandais; R. C. VI, 129, sqq.

2. In general the accent inherited from O. I. is retained. On weakening of the accent force new agencies became energetic in Munster, the O. I. laws being crossed by others with ensuing disturbance especially in the Dēsi.

I. A long vowel in an unaccented syllable induced strife between the stress and tone accent. The stress accent balances the tone accent or yields to it. In the latter event a detoned syllable before the new accent suffers the same reduction as the syllable after the ictus in O. I. *amadán* MUDẢN, *diombáidh* DM'ả, *coileán* KILẢN, or KL'ẢN, *taisbéan* SPẢN or SẢN *garrán* GRẢN *taréis* 'after', Tr'ēs.

Note. A long vowel cannot attract the accent over a heavy consonant group; *iompo* AUMPŌ, *caindleoir* Kainl'ōr'.

II. A heavy syllable unaccented has a like effect. *salach* SLảC, *beannacht* BN'ảCT, *marcach* MURKảC, *uireasba* rísə, *currach* KRaC. Where there is no such heavy consonant group the accent retains its legitimate position. *marcaig* MảRKIg:

Regarding the power of the heavy termination — *ach* to effect disturbance of the accent there is distinction between its uses as nominal and adjective suffix. As noun suffix it usually bears the accent, *bacach*, BUKảC, a lame one, a beggar but BảKUC lame. In pausa even the adjective termination may be accented, as *gortach* in the following example:

Fiodh-dún na gcrann
thá baile beag ann
gortach,
baile beag briste
is a thóin leis an uisge
is mná gan feiscint ann.

IV. Pro- and enclitic particles distinguish strong and weak forms according to stress and relation to the pausa position, *leo*, *leotha*, *lé*, *léithe*, *sin*, sin, SUN, and SN', *ann*, AUN and N'. *cé (= cia)* kē and ke, *mé* mē and mə *tú*, TU and Tə *tar*, har dar and Tṛ. *agus*, *is*.

V. *ag* in conjunction with pronouns is enclitic *agam* əGU'M *agat* əGU'T *aige aici* əGé, əkí, *againn* eGIŋ, *agaibh*, əGIV, *acu* əKU'. So *umam (= imb- i* to *u* before a labial) U'MM' and əMU'M. *orm* ÓRM' occassonally in pausa əRUM. *ni'n aon phoca orm* nīn E FÓKə rUM, Tom Lannon. Similarly *oiread* but *an dá oiread* N' Dả RU'D.

VI. Under the accent certain consonants doubled in auslaut of short-voweled syllables, or in position in inlaut, develope sounds which supplant the original root vowel or combine with it in the production of a new slurred diphthong. Failing the conditions e. g. when through an increase the syllabic division separates double consonants the primitive sound remains. Vid. § 3 sqq.

3. Pausa has a liking for long vowels and for the accent. Some verbs reserve a longer form for pausa. *Chondairc me é,* cnik me c̄, but in affirmative answer to a question cnikīs.

4. The irrational vowel ə is absorbed by a neighbouring coloured vowel in context. *Chuaidh sé isteach* hwū sē stàc, or cū, Two neighbouring irrational vowels contract to one.

5. Though unaccented syllables void their colour still they can but seldom be represented by the sign for the irrational grade from their liability to borrow a tint from following consonants especially labials and gutturals, including guttural ʟ. Cf. the ə from ɴ́ ᴍ́ becoming ᴜ under the accent in *ugam, sin, (san) is,* the unaccented form of *agus.*

The influence of a following consonant
on vowels.

§ 3,1. a) A slender consonant left the foregoing long vowel unchanged. Umlaut however often took the place of a short vowel. This process in complete in many instances, in others is has not yet begun.

b) A broad consonant developed a broad parasitic vowel which usualy umlauted a foregoing short root vowel.

2. a) A heavy consonant group lengthened, or far oftener, diphthongized a short root vowel under the accent. Thereby a great number of new slurred diphthongs have been developed. In Connaught and sometimes in Thomond lengthening only is apparent in those situations. Vid. Finck, *Wörterbuch der mundart der Araninseln.* Also in the East, in Man, tōlsə *fallsa,* surī *suirghe,* Thos. Kermode Bradda, Isle of Man. Vid. Strachan, *Zeit. f. Celt. Phil.* I, p. 54. In the Dēsi and

generally in Munster lengthening is found only after. *rr* or *r* in position. *fearr* fāʀ, *dóirse* pl. of *dorus*. Diphthongization is a special Munster characteristic and each subdivision is distinguished by a particular table of those new sounds.

b) Of the consonant groups producing this change some are permanently heavy (inlaut position) some capable of being lightened by the addition of a suffix. Sometimes in composition the accession of an element with consonant initial constitutes a heavy group. *Tein-loit* tᴇinlot, or tᴇilot T. G. (if from *teine* and not *teann), seandrui* sauɴᴅʀi *seandamh* sauɴᴅâv, *bainrioghan* ʙauʀīɴ (the writing *bain-* is only ornamental, the pronunciation comes from *ban-) antoil* auɴtĺ̦ Denn, *contabairt* ᴋauɴᴛuʀᴛ P. Walsh song, (Irregular accentuation, usually ᴋɴ′ᴛouᴛᴛ), *aungur i buairt an taelso*, Cat. 47. (But *seanduine* sáɴᴅinə *andóchus* aɴᴅōcis.) Here the first element was accented and formed an integral part of the word, if proclitic as in *andóchus* it lacked the essential of accent for constituting a heavy group in the sense of this rule. Cf. accent and proclitic groups *ionnsaighe* auɴsī but *ionntsamhail* iɴᴛūl. When a group capable of suffering such a process is lightened the syllable retains its original sound (radical or umlaut), after or before the accent it becomes the irrational vowel. *ceanfhionn* compounded of kauɴ and fyauɴ and pronounced kaɴɴ′ exemplifies both cases. *nín se ann*, auɴ, but *dailtin maith athá ann*, ɴ′.

c) A comparison of C. M. O. (Clare, 1780) and T. G. (Cork Kerry? same period) with the Dēsi usage revealed the following sound table. Oblique lines denote rare or exceptional intonation.

Dēsi	au			ou	î		ai	ᴇi	
Thom.	ū	\ou[1])	ā[2])	ou	i[3])	ī[4])	ī	ī	\
Des.	ū	ō		ou	ī	ē	ī	ī	ē

---

[1]) *meabhair: trom*, C. M. O. 6.   [2]) *fallsacht: fáidhchirt ib.* 7.   [3]) *gadhar, aighneas, faghairt taidhbhse*: i, *ib.*   [4]) Diphthongs arising from *mh, ch,* and *ll* as *rachad, doimhin, moill*: ī.

§ 4,1. au. This diphthong, which is produced in the accented position by the nasal *mh* surrounded by *a* vowels, or by the influence of double liquids (except *r* which caused lengthening) under the conditions set forth in the examples hereunder given, is of a strongly nasal character. It accompanies broad consonants and supplants all short root vowels except *o*[1]). From the fact that *ball* and *ceann* show this sound as well as *samha* sau, where its origin from a nasal is pretty clear (the *mh* being in fact an anusvâra sign), one may consider that a parasitic vowel developed nasal timbre before broad liquids and umlauted the root vowel to au. Whatever the order of the process the result from *mh* and double liquids is the same to-day. Hence it is found convenient to speak of an anusvâra developed by double liquids. So also where w and y are spoken of as developed by liquids there is merely question of identity of results, that is to say, liquids have behaved towards root vowels in the same way that has been observed in the case of w and y. The addition of a vowel even of a svarabhakti resolves an auslaut double group into elements belonging to separate syllables, such an anusvâra is not induced, and the original sound is retained. A liquid in inlaut in position produces an insoluble anusvâra. The groups producing the au diphthong are written thus: −*ng*, *ll*, *nn*, *m*═*mm* and *amha* in auslaut, -*nd*-, -*nl*-, -*bhr*-, -*mhr*-, -*mhn*- in inlaut. In all those cases au takes the place of the root vowel. *long*, *drong* and *Conn(?)* through influence of the guttural nasal make au though belonging to the ou diphthong.

Note. From the fact that an addition breaks up a heavy liquid group there arises the phenomenon of many words having a long vowel or diphthong as monosyllables but showing a short vowel i. e. preserving the original length as dissyllables. *snaidm* sᴎím, pl. sᴎâᴍɴ´ə, as if from *snam-mana*, *greimm* gr´îm,

---

[1]) o+w from *ll*, *mm*, *nn* in auslaut or position or from *dh*, *gh*, *mh*, in inlaut became ou. § 5,1.

pl. gramɴ'ə, *barr* Bǎʀ but *barra* Bàʀə, *gleann* glauɴ gen. glaɴə. Hence sometimes a dissyllable took a short vowel by analogy. *tabhair dam* ᴛᴏur ᴅuᴍ has ᴛᴏᴛᴍ' as unaccented form (shortened from a form ᴛōr ᴅuᴍ) *comhthrom* from ᴋōrhᴍ' or ᴋᴏurhᴍ' to ᴋᴏrhᴍ'. Also in *mise, tusa, ise*, we must either say that the unaccented short form of the pronoun was chosen as stem for the emphatic increase or else *mé tú* and other such monosyllables have become lengthened by some unknown law. *h* written *-th-* has a power of absorbing y. Thus the termination *-the* of the ppp. of *-ighim* verbs absorbs *-gh-*, there is no coalescing of y with *i* and hence no lengthening. *malluighim* ᴍàlīm, *malluighthe* ᴍàləhə. Even so *-th-* in the *t, d,* declension or what has become such by late analogy. The word for house is declined *ti, tighe, tig* which is pronounced ti tī tig pl. tihə. *dlighedh* glī pl. glitə: *fuirean* Ren. 69, p. 34, *troigh : dlaoi* therefore ᴛʀĪ, pl. ᴛʀɪhə, dim. ᴛʀɪhīn. *croidhe* ᴋʀĪ pl. ᴋʀɪhə Denn, 66. *deachmhadh* 'tithe' dahŪ pl dáhəhə.

Examples:

ball	ʙaul	*cionta*	ᴋauɴᴛə
fionn	fyauɴ	*cantla*	ᴋauɴᴛʟə
ann	auɴ	*drantán*	ᴅʀauɴᴛǎɴ
seanda	sauɴᴅə	*liomsa*	lauᴍsə
fallsa	ꜰauʟsə	*sganradh*	sɢauʀə
dream	drauᴍ	*branradh*	ʙʀauʀə
cam	ᴋauᴍ	*splanc*	sᴘʟauɴᴋ
prionsa	pr'auɴsə	*ranclach*	ʀauɴᴋʟuᴄ
lansa	ʟauɴsə	*seang*	sauɴ
damsa	ᴅauᴍsə	*young*	yauɴ
glamhsán	ɢʟausaɴ	*treabhladh*	trauʟə
		iomchur	auᴍᴘʀ'
		stumpa	sᴛauᴍᴘə
		blandar	ʙʟauɴᴅʀ'
		banlamh	ʙauʟuv
		meabhrach	mauʀuᴄ

gamhna	ɢaunə
samha	sawə
romhat-sa	ʀautsə
seandrui	sauɴdʀī
coinsios	ĸaunsīs
frainsios	ꜰʀaunsīs

In the last two examples the foreign slender group *-ns-* was regarded as broad, otherwise only broad groups produce this sound.

2. Unaccented au from *amh*=ᴜ. *damhsa* ᴅūse, *amhail* aul but *fearamhail* faʀūil, *modhamhail* by contraction ᴍoul. Near the accent anusvâra is not developed and the syllable is treated like all other detoned cases. *druimfhionn*, ᴅʀɪᴍɴ'.

3. *ll, mm, nn (rr)* being resolved on receiving a vocalic suffix the primitive vowel is retained. *ball* ʙaul but *ar balla chrith* er ʙal'ə cr'i *gleann* gen. *gleanna* glaɴə. *cam* from *cambo-*=*camm*, *cama chlis* ĸàᴍəclis, *crann*, but *crannaibh* ĸʀaɴiv, *mionn* pl.

4. *rr* in auslaut and *r* in position lengthen. *ferr* fāʀ and fåʀ *gearr* gāʀ, gåʀ. In position *dorus* but *dóirse, sùgradh*. When the combination is broken the original sound remains. *is fearra dhuit* is faʀə ʏot or ʏet, *máire ghearra* ᴍår'ə ʏaʀə, *sùgradh* sᴜɢr':

§ 5,1. ou. With broad consonants. Accented short *o*+w arising from *ll, mm, nn*, in auslaut, arising from *gh, dh, mh*, before a voiced element, and short *a* or *o*+w from *bh* before a voiced element, combine to ou.

coll	ĸoul	*bodhar*	ʙouʀ
bonn	ʙouɴ	*fodhb*	ꜰouʙ
drom	ᴅʀoum	*foghlach*	ꜰouluc
rogha	ʀowə	*tomhas*	towis
labhairt	ʟouʀt		
seabhac	souĸ		
abhac	ouĸ		
Ultach	oultuc		

as if *oll-* Vid. derivation of *Uladh* in poem quoted Manners and Customs Vol. I, p. 8.

2. On receiving a vocalic increase *ll, mm, nn,* are separated and the original sound is not disturbed. *Croma* KROMə, *droma* DROMə. In *drong long* the anusvâra induced by a nasal has umlauted *o* to au, vid. § 4,1.

3. Unaccented=U. *comharba* KUərBə, *Concobhar* KUNŪR. Originally unaccented prefixes retain their close sound even when they become accented. Thus the first syllables of *comhursa* and of *commaith* are sounded alike although *comh* in *comhursa* is now accented while the *com* of *commaith* is still proclitic. So *iongantach* Uəɴᴛᴜᴄ, though now accented on the first syllable. The pronunciation auɴᴛᴜᴄ T. L. in due to late accent working.

4. In Desmond sometimes=ō. *comhachtach : dóchus* T. G. 4, *rompa : dóirse,* T. G. 5, *sómpla* T. G. 13, Dēsi saᴜᴍᴘʟə. *ómbrach* T. G. 23, Dēsi auᴍʀʹᴜᴄ.

5. In the Dēsi pronouns and prepositional pronouns are usually exempt from the action of those change laws but *liom* lᴜᴍ often lauᴍ in songs, *ionta* iɴᴛə and auɴᴛə *romhat* ʀauᴛ, or ʀawət. In Thomond and Desmond they usually conform *sinn* sīɴ.

§ 6,1. i, with broad and slender consonants. y from *gh, dh,* broad and slender and from *mh, bh* slender, and the y developed by slender *ll* in auslaut or slender *l* in position changed a short root vowel (except *ei=e*) to i under the accent. Groups denoting the presence of this sound are usually written: *-ll, -ls-, -llf(=llh), -lt-, -ghn-, dhbh-, gh, mh, bh,* slender and *gh, dh,* flanked by *a* vowels or preceded by *a* and followed by a voiced element. The change of *gh dh* to y in these broad positions appears exceptional. A parallel change of broad *bh* to y is found in words like *diabhal* dīyʟ and dyouʟ, *riabhach* rīyᴜᴄ, where one may consider *bh* as assimilated to the following consonant and y as a glide from *i*. In some instances *dh* was originally slender. *radharc* from *derc, adharc* a horn, *adercéne* gl. corniculum, Z^2 274. After working of the accent law the vowel of the second syllable was reduced to ə and written *a* the consonant remaining slender. *adhmad* 'brennholz' Finck, and so connected with *adannadh* .|. *adn´ad* (*ann*=n´ unaccented, cf. *anmanna*) *ayn* Cat. 3,9, and *adhradh* like Lat. *ad-orare, ayrim* Cat. 9 appear to have had broad *dh.* So *Tadhg.* Broad *dh*=y in accented auslaut in Manx, MU GRai-I is mis´ *mo ghrádh is mise,* Thos Kermode, Bradda Isle of Man. In unaccented position also *dh* broad=ẏ, ex. *-adha* as nom. pl. ending=I, *curada* KURI. But also=*w. iomdha* MŪ (but the prep. *imb-* was unaccented here), *bunadhas* BN´Ūs.

Examples:

2.
cill	cil	*luighead*	Lîəᴅ
millteach	mîltuᴄ	*laighean*	Lîəɴ
soillse	silsə	*foighnne*	Fiꞑə
coill-te	ᴋil-tə	*aibhne*	iənə .
tuill-fead	ᴛil-heᴅ	*doimhin*	ᴅiən
aighneas	înis	*maidhm*	Mîəm
saidhbhreas	sivris	*tadg*	ᴛîɢ
	badhb	Bîʙ	
	cladhaire	ᴋLîər´ə	
	adhmad	îMəᴅ	
	faghairt	Fîərt	
	bfuighbhinn	vîꞑ	
	coigthigheas	ᴋîkīs	

3. *d* is assimilated (?) to *g, b, m, n,* in *tadg, badhb, snaidm, adnadh,* etc. Hence it should not be dotted in such situations. If one may depend on the present pronunciation of *coigthigheas* ᴋîkīs, kaikīs (?) Finck, *coig-* must be separated from O. I. *cóic* because î proceeds from a short vowel. Neither has it to do with Lat. *quinque* because of ᴋ.

4. Unaccented=ī, *curadha* KURī from y;=Ū from w. Vid. *bunadhas* above. On separating double liquids y is not developed and the primitive vowel is preserved, *cill,* gen. *cille* kilə. It

falls away in detoned position and the syllable suffers the usual reduction. *eochaill* ōh̥l 'Youghal' (!) *ch*=h.

§ 7,1. ai, with slender consonants. y developed from *mm, nn,* (=ŋ) or *mm ng* in position contracts with *i* of root under the accent to ai. The groups are written: *-imm, -inn, -imch, -int,* etc., *cuimhin* which should belong to the foregoing is umlauted to ai by nasal influence.

binn	baiŋ	*simplidhe*	saimplī
clainn	ᴋʟaiŋ	*timchioll*	haimpl̥
cruinn	ᴋʀaiŋ	*ling*	laiŋ
impreamhail	aimpr̥Ūəl	*linn*	laiŋ
intinn	aintiŋ	*trinseach*	trainsᴜc
rince	ʀaiŋkə	*saint*	saint
im	aim	*slim*	slaim
suim	saim	*rinn*	ʀaiŋ

grinn	. gr′aiŋ
crainn	ᴋʀaiŋ
aoinnech	aiɲe
coint (Fr. Conte)	ᴋaint, ᴋoint,
caincin a retroussé nose,	ᴋaiɲcīn
cingcis	kaiŋkīs
cuimhin .	ᴋain

2. A vocalic addition induces separation of double consonant groups and preserves the root sound. *binn* but *binne* biŋə. Nouns of the *o* declension ending in *mm, nn,* belong here in gen. voc. sing. and nom. pl. *ceann,* gen. *cinn* kaiŋ. Pronouns and prepositional pronoun compounds are excepted *sinn* siŋ, Desmond, Thomond, sīŋ. *linn* with us, liŋ but *linn* a pool, laiŋ. § 5,5.

3. Nasal influence cancelled by breaking pausa=I *cuimhneamh* ᴋīnᴜv=regular accented intonation Thom. Des. § 3,2 c.

Near the accent=ə or ŋ, *druimfhinn* ᴅʀimŋ.

§ 8,1. ᴇi with slender consonants. A variant of the preceding. Following *ei* (=e) the y from *gh mh bh dh* slender or developed by double final liquids or single in position changes *ei* into ᴇɪ. A following *r* produces the same Diphthong. *eirghe*, ᴇɪr'ī *teir* tᴇir', also Eng. 'pair', 'stairs' become ᴘᴇir'ɔ, sᴛᴇir'ɔ through *r* influence.

2. The combination are *-eimm*, *-eill*, *-eir*, *einn*, *-eigh-*, *-eidh-*, *-eibh+*a liquid.

Examples:

feill	fᴇil	*deigh-*	dᴇi
sgeimhle	sɢᴇilɔ	*teimhill*	tᴇil
geibhleach	ɢᴇiluc	*leigh-te*	lᴇi-tɔ
teighim	tᴇim	*eibhlin*	ᴇilīn
feighil	fᴇil	*beinn*	beiŋ

meidbh	mᴇiv
eisteacht	ᴇistucт
meadhar	mᴇir'
greadhan	gr'ᴇiŋ

eisteacht : eula, Dunne from *éitsecht*, Dēsi ᴇis —. *meadhar greadhan* are also written *meidhir, greidhean. medar*, W.

3. Vocalic initial suffix divides auslaut group and the old sound remains *greimm*, but *greamma*=gr'amɔ.

4. Some few words belonging either to ai or î have made a separate class with oi vowel. They are:

raint	ʀoint	*suim*	soim
maigistear	ᴍoistr̩	*muintear*	ᴍointr̩
maigistreás	ᴍoistr̩ås	*mainseach*	ᴍoinsuc
timcioll	hoimpʟ'	*baintreabach*	ʙointruc

biombéil	boimbēl
caill	ᴋoil
scim	skoim
gaillseach	ɢoilsuc

Some of those are always pronounced oi, *maigistear -ás, mainseach, baintreabhach, biombéil (bíoma* Eng. 'beam'?) others oi, ai, î, *suim, scim, gaillseach*

§ 9. Hereunder the symbols of the historical orthography will be taken in order and an attempt made to set forth the sounds they represent. The vowel-signs whether representing vowels, diphthongs, (real or merely timbre digraphs) or umlauted sounds will be treated in some detail. This has been considered necessary for the reading of an assonantal system of verse, particularly as the sounds intended to be conveyed cannot in many cases be even divined from the common orthography. The consonant signs will receive shorter treatment because their value is mostly sufficiently evident. Special regard then will be had only for those cases when the symbols no longer represent the sound, for example in occurrences of assimilation and combination of vowels with remains of affected consonants to contraction or diphthongization. For a treatment of the liquids I refer to the minute and exhaustive discussion in Dr Pedersen's work *Aspirationen i Irsk*, Leipzig 1897.

Of the Vowels.

1. Short vowels.

§ 10,1. *a.* Usually = à. *cad* ᴋâᴅ, *cara* ᴋârɔ (pl. ᴋăr'dɔ, lengthening from *r* in position) *ca bhfios duil?* ᴋàvis ᴅoᴛ, or ᴅcᴛ, **can fiss duit, ca shoin* when? ᴋá'cꞩ and ᴋác'in, also[1]) ᴋăcꞩ. *ca'l se?* ᴋàl sē and ᴋăl sē, *talamh* ᴛàluv, gen. *talmhan* ᴛàlᴜɴ, *mac* ᴍăᴋ, and so for nearly all the occurrences. It is labial or guttural umlaut from *a*.

2. In some words however *a* in accented syllable in a) anlaut or b) preceded by a consonant that is neither a labial nor a guttural (including guttural ʟ) = a. a) *anfa* aɴɔhɔ, *admháil* aᴅɔwằl, (followed by *n, m* = à; *anam, an,* 'stay', *amus,* but *an-* 'very' aɴɔ-, *anfa* a great blowing,) *arm* aʀᴍ',

---

[1]) Here we have broad *sh* ∠ C by some change process obscure to me. The parallel charge of slender *sh* ∠ c, through *hy* as explained by Pedersen is quite common and may even be heard in English as spoken in Ireland. F. g. *human* cᴜᴍɴ'.

argain arǝɢɴ. *as* prep. with its pronominal combinations *asam*, etc. *athair* ahṛ' *asal*, *athchuinge* ᴀᴄɪɴe. b) *tarraingt* ᴛᴀʀɪɴt, and so *tarbh*, *dar* 'by' in oaths, *tar* har and compounds, *tarra* 'tar' *dara* ᴛᴀʀǝ (the spelling descends from *ind aile*, the pronunciation from *int aile*). *Crathach* one named M'Grath, *tarbha* ᴛᴀʀǝғǝ, *lagan*, but *nar thagaidh tú* hàɢǝ, *tafant* ᴛᴀғɴt (ᴛᴀhɴ', Com. song) *dafhichid* ᴅᴀhɪᴅ, *stad* 'stop' *tannidhe*,[1] *dadam*, middle Irish *ttada*, *apstallaibh* T. G. 38. *Sasanach*. Loan words often kept a, hence *hata* 'a hat', *patan* a patent, a title deed, *a bpront Patain a meamram táirne (= tarraingthe)* Grace song. O'Neill. This word took an infixed *r* and appears later in the puzzling phrase *cur i bpratain* to commit to writing, relate deeds. Also *agam*, etc. when for metrical reasons accented on the first syllable.

3. *a* is usually preserved before a softened consonant This of course is written *ai* q. v. In *gabhtha* = ᴏ : *obair* C. M. O. Here participle from *gabail* = ɢōʟ shortened before *th*.

4. *a* becomes ǝ in unaccented position. This is liable to take a u colour from guttural or labial (u shaped) consonants. *gach aon* ɢᴜ hᴇɴ, *casog*, 'cassock' ᴋᴜsōɢ, So *a* = ǝ the svarabhakti or vowel sign from ļ ṃ ṇ on receiving the secondary accent becomes ᴜ under the same conditions : *ansan* ɴ'sᴜ'ɴ, but *athà san* hǎ sɴ', *agam* ǝɢᴜ'ᴍ, *agat* ǝɢᴜ'ᴛ, *dam* ᴅᴜᴍ, : *dubh* P. P. 226, (not a shortened form of *damhsa* ᴅōsǝ but ᴅᴍ' accented).

5. *a* = nul through contraction of an unaccented syllable. *salach* sʟàᴄ, *tar éis* ᴛʀēs. *a* = ǝ in anlaut or auslaut is absorbed by a neighbouring coloured vowel. *bhi cailíni oga ann* ᴠī ᴋᴀʟīɴī ōɢ ᴀᴜɴ, *chuaidh se asteach (isteach)* ɢŪ, or ғŪ, or hwŪ sē stàᴄ, *rachadsa ann* ʀɪᴅs ᴀᴜɴ.

6. *a* makes a number of slurred diphthongs with the semi-vocalic relics of affected consonants, by contraction, and

---

[1] Of the O. I. forms *tana, tanide* the latter only is used in the Dēsi.

by occurring in heavy syllables under the accent. With y
from *gh, dh,* etc., to î and *y* glide to following vowel. *aghaidh,*
iyə, *radharc* RÎRK, RÎYəRK, *Tadg* TÎG, *cadhata*(?) *maith dhe
cheapaire* KÎTE MÀ yə (gə) cápɽə a good piece of bread and
butter. *ladb, badb, snaidm, adradh,* etc. vid § 6. But *rachainn* :
talamh D. R. 37, RÁCIŊ = western intonation, *gadhar* : *séimh*
P. P. 140 is a non-Munster pronunciation.

7. *a* unaccented with *w* from *bh, dh, gh, mh,* to Ū.
dearmad daRŪD. daRŪDIŊ 1st per. imperfect, daRŌTIŊ 1st per.
conditional. Here ō is adopted from *-ighim* verb futures,
the *t* is a result of *df* = *dh* combination. S. R. So *-amha*
or liquid + *mha, bha,* etc. in unaccented position. *marbhadh*
MÀRŪ, *marrugh* Cass. 141, i. e. as if *marughadh* hence this
and a good many other verbs have gone over to the *-ighim*
conjugation, fut. MÁRŌəD. *berbhadh* berŪ, *a g-cualabhar* ə
GŪLŪWɽ S. R. *iomdha* MŪ. *maiteamhnas, mohunas* Cat. 23,
contabhairt KN'TŪrt, but once KAUNTŪrt Walsh (song) through
exceptional stressing of the unaccented preposition. *bunadhas*
BN'Ūs, *ladamhas* LÀDŪs, *damhsa* DŪsə, *greannmhar* gr'aNŪr
deachmhadh dahŪ, pl. dahəhə, *ceatramhadh* kāRhŪ. By
analogy KŪGŪ (ū after a slender consonant) tr'ĪGŪ from
mixed analogy of the numerals for 'four' and 'five'. *Talmhan,*
(gen.) TÀLŪN, *beulmhach* biàLŪc a bridle bit, *feolmhach* fYŌLŪc,
murrghach MURŪc, *searbh* SARUV but *searbhas* SARŪs, *cearbhach*
(from *carbhas* 'carouse'?) KARŪc a drinking man, by popular
etymology restricted to 'card'-player. Also endings in
— *amhail* Ūl i. e, Ūwəl. Cf. parallel of slender consonant signs
to y after liquids, *suirghe* SURĪ, *inghean* inĪəN.

8. *a* = ắ, ā, before *rr* and *r* in position. *bárr* Bằʀ,
barr W. *cáirde* pl. of *cara, táirne* sometimes used for *tarraingthe.* This lengthening influence of *r* extends to all the short
vowels but is not always energetic. Thus *airc* ark. The
svarabhakti vowel from *r* breaks position and in such examples
as *arbhar* aRŪR *bh* is vocalic. When the syllable is lightened

by addition of a vowel *rr* in auslaut is sundered and there is no lengthening:

an chroch árd dá cur 'na seasamh

is an buinneán aorach dul 'na barra. .|. ʙârə

9. *a* in -*adh* verbal noun ending = ə. After a liquid + *mh, bh, dh,* = ʊ̄. Vid § 10,7. *bualadh* ʙᴜ̄əʟə but *berbhadh* ʙᴇʀᴜ̄. Unaccented -*adh* in auslaut therefore gives up its *j* or *w*. *adh*- accented = î vid. § 6. In Desmond often = ᴇ *madhmach:laochradh* T. G. 12. Accented auslaut = ī *cladh* ᴋʟĪ, so -*aidh* in *ceadfaidh, chadfuy,* Cat. O. I. *cétbaid* with accent on first syllable.

10. *adh, agh* as accented monosyllables = å, *ádh, ágh, rádh. adhbhar* åvʀ́ S. R. usually ᴏᴜʀ. Here there is old lengthening.

11. *a* = å by contraction *Adhamh* åv, T. G. 9. 19. *Aav,* Cat. 31. From *namha* ɴå. *baile atha cliath,* ʙʟᴀ̄ᴋʟī, Here contraction to ā and exceptional loss of *th*. This, the pronunciation of a place name, came from outside. Intervocalic *th* is not lost in the Dēsi.

12. *a* with *w* of *bh* under accent to ᴏᴜ. *cabair* ᴋᴏᴜʀ, *tabhair.* Vid. § 5,1. Often the older stage åv is kept in songs. *labairt* ʟâvṛt, *cabair : chailleamar* T. G. Therefore ᴋâvʀ.

13. *a* which with anusvâra *mh,* or induced from-*nn, -ll,* etc. = ᴀᴜ. *lann, mall* ʟᴀᴜɴ ᴍᴀᴜʟ, (Vid. § 4,1) sometimes preserves non Dēsi forms in songs; *do réir m' amhrais* ʀᴄ̄ʀ ᴍᴏᴜʀɪs, M. song. *amhras : cabair* Brett apud Dunne. When anusvâra is not induced the root sound remains. *Cama chlis* ᴋᴀ̆ᴍə clis.

14. *a* = ᴏ in ᴛᴏʀᴍ́ .|. *tabhair dham* unaccented. Here *tabhair* once = ᴛᴏ̄r is shortened by the addition of a syllable. Cf. -*the* plurals *rí, rithe* ʀɪʜə : *imthig,* C. M. O. In such cases resolved pausa shortened on the analogy of the treatment of real pausa lengthenings.

15. *a* sometimes arises in plurals and adjectives restored from obscured monosyllables by the breaking of a heavy con-

sonant group or through the resolution of pausa. *dlaoi* (?) *dlathach*, adj. *na dlathaibh* P. P. 178: *snamach=snam-mach* from *snaid-m* sNîm, *claoi* .|. *cladh*, KLĪ, *klai* Finck, pl. KLahucə. Sometimes an *a* appears by false analogy *rig* pl. *rathacha* D. R. 98. 'arms'.

16. *a*=ō (from aw) in *gabháilt*=going, moving, causing to move. *gabhailt na gaoithe* : ō C. M. O. 17, *gabhail leo*: *pósadh* ib. 10. This verbal noun has three pronunciations, GUvål, as much hay as one may carry in the arms, GWålt, the usual verbal noun, and Gōlt, |going. So *adhbhar* : Sport in Dunne, Carrick-Shock song. Also *namhad* : ō ib. A lengthening from *r* in position, *thôrsta* 3 sing. fem. prep. pron. from *tar*, M. song. So 3 pl. *torsta* : *tabairt* : *óg* P. P. 136.

E, e,

§ 11,1. E is the ē sound after a broad consonant. It is usually written *ao* sometimes *ae* and represents the O. I. diphthongs *oi*, *oe*, *ai*.

2. e appears only in auslaut as it forms a digraph with the timbre index of a following consonant. Before a broad consonant written *ea* and sounded a, before a slender consonant *ei* and sounded e.

3. *e* accented=e. *aige*, əge. For *ag* O. I. *oc* the pronominal form masculine is used.

4. *e* unaccented=ə. *báidhte* Båtə, *tar-se* TARSə, *léthhe* lēhə.

5. *e* unaccented with *y* from *gh*, *mh*, *ch* after liquids and *t* contracts to ī. This long vowel then drew to itself the accent. § 2,2 I. *eirghe* EIr'ī', *gainmhe* Ganī', *suirghe* sur'ī', *inghean*, inī'əN, *doilgheas* Dolī's, *coitcheann* Kotīən, *faitcheas* Fåtī's, *dairghe* Darī', *comairce* KUMṛī'. Cf. § 10,7.

§ 12,1, I, i.

I is the *i* sound after a broad consonant. It is usually written *ui*, *oi*, and in often the i umlaut of a broad vowel, *fuil coille* FWIl KIl'ə.

2. *i*=i. *mise*, *ithe*, *binib* T. G. 'Venom' .|. savage spite.

nithe nihə pl. of *ni*, *dlighthe* glibə pl. of *dligh* glī, *tighthe*
tihə pl. of *ti* ti, gen. *tighe* tī, dat. (nom.) *tig* tig, *imthighthe*
iməhe and mihə, *linn* liŋ prep. pron. *dliher* Cat. 25 *dlighear*,
Koatynge.

3. *i* by contraction with *y* remains of consonants = ī.
nimhe gen. of *nimh* poison : *ti* T. G. 10, nī, *sligh* slī, *brigh*
br′ī, *tighe* tī. So the *-ighim* terminations of verbs *malluighim*
ᴍâlīm. Some tenses of some simple verbs have gone over to
the denominative conjugation *ni feicighim ann é*, *condaircigheas*
cɴɪkīs I saw (in pausa), *sint* Com. song, a Desmond lengthening,
the Dēsi form is *sinmhuint* sinŪnt form *seinm*, *Brighid* brīd.

4. *i* = ai under the conditions mentioned § 7,1. *tinn*
taiŋ, *sinsear* sainsᵣ, *simplidhe* saimplī, *inis* inis, but *inse* ainsə,
Sceichin na rince sgehīn ɴəʀaiŋkə (′ring′) The dance Bush
(a place-name) *mo ghrinn croidhe* ᴍu yr′aiŋ ᴋʀɪ.

Thom. and Des. examples: *binn* : *sios* Dunne, *sinsear* :
sioda Sheehy, *cinte* : *rimhthear* Sheehy, *tinte* : *fiochmhar* C. M. O.
pl. of *teine*. Prep. pron. *linn* līŋ, pron. *sinn* sīŋ.

5. *i* under conditions described § 6, 1, = î. *cill* kil. Latin
cella should give *ceall* but place-names being mostly spoken
of with reference to motion to or from the dative finally
became nominative. gen. *cille* kilə, *coill* ᴋɪl, gen. ᴋɪlə, *milis*
milis, but *milse* mîlsə, *mill* mîl, but *milleadh* milə.

6. *i* quite rarely under the accent = u. *go dtuga tú sin*
dúinn ɢu ᴅuɢə ᴛŪ su′ɴ ᴅŪəŋ, Crowley, *ca'l a leath-cheann sin*
ᴋâl ə lacauɴ su′n. Here there is an upraising of the vocalic
portion of *n* sonans. Cf. *agam*. In *rith*, *rioch* ʀuc, *i* is umlauted by the *u* thrown off by the guttural spirans.

7. *i* in a couple of words under accent = e. *file* felə
(Cf. *veleda*) *bile*, belə, *air beille breagh geagach*, Fitzgerald
Poems, Ren. P. 65. Also *go bhfelleadh* = *go bhfilleadh*, Dunne.
This is against the tendency to exchange *e* for *i* under the
accent, cf. Eng. 'pen,' 'men' pronounced pin min in Ireland.
There is an etymological reason for felə, the e in *bile* is

obscure to me, and *bhfelleadh* is probably due to the influence of *file*.

8. *i* unaccented＝ə. *licighthe* 'likəhə, schlank', *sillice búrca* siləke ʙᴜ̄ʀᴋə, a snail.

9. *i* unaccented often further reduced to nul. This occurs when unaccented *i* in context is absorbed by a neighbouring coloured vowel *mā's* ＝ *mā is*, *lionta bpiantai*b C. M. O. for *lionta i bp.* or when it immediately precedes the accent, *dubhairt leis 'mtheacht leis féin*, for *imtheacht*, *'dir* ＝ *idir*.

10. *i* ＝ ē in *gidh* pronounced kē, the writing from O. I. *cid* the pronunciation from *ced*.

§ 13,1. *o*＝ᴏ. *tabhair do é* ᴅᴏ, *innso* ənsᴏ, *gob* ɢᴏʙ, *gotha*, gen. of *guth*, ɢᴏhə, *trosgadh* ᴛʀᴏsɢə, *sop* sᴏᴘ, gen. sip.

2. *o* unaccented＝ə. *dob' oige* ᴅᴇ ʙᴏ̄ɢə, *tosach* ᴛɪsᴀ̀ᴄ. Here e to i through influence of *s*. So *u* in *dorus* ᴅᴏʀɪs. *ar mo chean*‾ er ᴍᴜ ᴄᴀᴜɴ, ə to ᴜ on account of the labial.

4. *o* accented in a few instances＝a. *bhi 'nnso 'guinn* ᴠɪ̄ɴ'sá ɢɪɴ, *thá sé 'nnso* hǎ sēɴ' sá.

4. *o* in *os* prep. unaccented＝a. *os ceann, os comhair, os a dhiaidh*＝as.

5. *o* in accented position usually near *n*＝ᴜ (Cf. the parallel change of ó to ᴜ̄ near *n* § 18,3), *sonus* sᴜɴɪs, *donus* ᴅᴜɴɪs, *moch* written *much*, Dunne. *gonta* ɢᴜɴᴛə, *Donnchadh* ᴅᴜɴəᴄə D. R. 95, (not ᴅᴏᴜɴ- because of svarabhakti vowel) *cnoc* ᴋɴᴜᴋ, *lonradh, lunra*, Cat. 9.

6. *o* with i umlaut rarely＝e. Vid. *oi*. *orrainn* erɪɴ Crowley in fr'aɢꞬ s ꜰᴏ̄ʀ erɪɴ.|. *freagair is fóir orrarin*. Also *orraibh* eriv.

7. *o* before the accent＝nul. *mo fhear* ᴍᴀʀ, *colúir* ᴋʟᴜ̄r, Com. song, *a Lifer*＝Oliver, Oliver Grace song 1687 apud O'Neil. *scolúire* sɢʟᴀ̌r'ə, D. R. 86.

8. *o*＝ᴜ̄ in combination with consonant remains in unaccented position (accented＝ou § 5,1). The prep. *con-* as element of a noun compound was proclitic and though now

stressed it mostly preserves the intonation of its unaccented period. *comharsa* ᴋᴜəʀsə, *comhnaidhe* ᴋᴜɴɪ *chommór leis* ᴄᴜᴍᴜʀ, *comhartha* ᴋᴜəʀʜə, *comhrac* ᴋᴜʀᴜᴋ, ᴋᴜᴍål Eng. ʻcombing', *comhradh* ᴋᴜʀå also ᴋᴏᴜ- and ᴋōʀå.

9. *o* which should become ou=ᴜ in *foghmhar* ꜰᴜʀ or ꜰᴜwəʀ perhaps through the labial. Pronounced also ꜰōʀ.

10. *o* which should also become ou=ᴜ through nasal influence in *foghantach* ꜰᴜəɴᴛᴜᴄ, Bob, *foghnamh* ꜰᴜɴᴜᴠ, M. *domhnall* ᴅᴜɴʟ', *congnamh* ᴋᴜɴᴜᴠ *cunah*, Cass. 145. *congarach : dlúith* Sheehy. So also in *lonradh : smúit* Sheehy, ʟᴜʀə Com. song, *lunra* Cat. 9 without assimilation of *n* to *r*. *lunrach (lonrach) : lúith-glic* Sheehy.

11. *o* which should become ᴜ occasionally=ō the regular Thom. Des. development. *comhrádh : leo* O'Shea from *corrádh* with lengthening before *rr* through remembrance of long ᴜ, for *rr* does not lengthen before a vocalic opening. *romham* (rather *reomham* Dēsi and Thom.) Des. : *sró'l*, T. G. *aball-ghort* ᴏᴜʟōʀᴛ lengthened before *r* in position. Here the word *abhall* an apple-tree has been preserved *(ubhall* ᴜəʟ an apple) but is pronounced ᴏᴜʟ- instead of ᴀᴜʟ because the accent has been drawn by ō. *foghmhar, four,* Cass. 145, like Eng. ʻfour' *comharsa, coarsa,* Cass. 145, like Eng. ʻcourse' ᴋōʀs, *comháircamh* ᴋōʀᴜᴠ, Bob. These occurrences in the Dēsi are probably loans from Thom. Des. where the accent does not distinguish‿ō and ou forms. The possibility arose from intercourse with labouring men from Des. who frequented the Dēsi in the last and beginning of the present century. *rómpa : dórse* T. G. *omhanda : scolta* P. P. 120.

12. *o* = ō before *r* in position. *dóirse* pl. of *dorus*. *compórdach* T. G. 2, from Eng. ʻcomfort', *compórd* T. G. 10.

13. *o* under the conditions mentioned § 5, 1, = ou. *foghluim* ꜰᴏᴜʟṃ *modhamhail* ᴍᴏᴜl with contraction, *comrádh* ᴋᴏᴜʀå, Crowley (*com* accented) *a leithéidi eile do chomhrádh bó is do bhi 'cu*, such an alarm as they made! So *fonn, poll, trom*,

= *tromm*, *comm* M. song, *com singeán* = *comm* a lake name in the Dēsi, *côll, fogha, tomhas, bodhradh*, ᴋᴏᴜʟʜᴜᴄ condit. of *codlaim*, Bob. ᴅᴏᴜᴛ = Eng. 'doubt' and ʙᴏᴜᴛᴌ 'sometimes' is pl. of Eng. 'bout'.

14. *o* which should be ᴏᴜ irregularly = au through nasal influence. *long* ʟᴀᴜɴ, *contabhaurt* ᴋᴀᴜɴᴛᴜ̄rt, Walsh (once in song) *contae* ᴋᴀᴜɴᴛᴇ, Eng. 'county', *anonn* ᴐɴᴀᴜɴ, *drong* ᴅʀᴀᴜɴ, pl. ᴅʀᴏɴᴐ. *romham* and *romhut* should be *reamham* etc. and so the pronunciation goes back to the older form. „*buil fhios agat cad athá do bhriste (d) á rádh*"? was a question asked of a knock-kneed man with a new corduroy breeches „*ni fheadar.*" „*Rachad-sa reamhat-sa, rachad-sa reamhad-sa!*" ʀɪᴅsᴐ ʀᴀᴜᴛsᴐ.

§ 14, 1. *u* accented = ᴜ. *muc, do chur, muthall* a 'mop' of hair on the head, *putal* a cherub, *futa-fata*, hurry, confusion, *sruth* sʀᴜᴄ, gen. *srotha* sʀᴏʜᴐ, *musdar*, P. P. 228, *bun, puthaoil* belching. The ending variously written -*aoil*, -*ighil* is attached to all verbal nouns denoting soughing or sibilation. *feadaighil* whistling, *glugaoil* the escaping of air bubbles on steeping clothes in water, (*glugar* an addled egg from its sound when shaken) *plubaoil* the rattling of water in shoes, etc.

2. *u* unaccented = ᴐ. *cróchur* ᴋʀᴏ̄ᴄʀ̣ a bier, *mo ghrádh fém tu* ᴛᴐ, *a dtuigeann tú me*, ᴐᴅɪɢṇ ᴛᴐ mᴐ, *solus* sᴏʟɪs from *s*. By metathesis *turas* ᴛʀᴜs D. R. 106.

3. *u* before the accent = nul. *currach* ᴋʀᴀ̂ᴄ, *fuláir* (= *furáil*) ꜰʟᴀ̊r'.

4. *u* = ᴜ̄ by contraction. *rugadh* to *rughag* and ʀᴜ̄ɢ, heard once from a man of the Dēsi (S. R.) while reciting a poem of Thom. origin. *cubhra?* ᴋᴜ̄ʀʜᴐ, *subhach* sᴜ̄ᴄ, T. G. 57, *dubhach* ᴅᴜ̄ᴄ. Before *r* in position *súgradh* : *dúil* M. song, ordinarily sᴜᴄʀ̣ in the Dēsi; ʀ̣ opened position. *lughaide* ʟɪᴐᴅᴐ, Dēsi. Terminations of verbal nouns in -*ughadh* ᴜ̄- or ᴜ̄wᴐ, as the sound stream is continued until the lips regain the natural position. *umhail* ᴜ̄ᴐl, *ul* Cat. 22, *uval* ib. 27. The phonetic

system of Cat. depended a good deal on the word-picture. *Mumha* MŪ, gen. *Mumhan* MŪƏN. *chum* .ǀ. *dochum* is pronounced CUN and CŪN according to the accent. *dar ndoigh thá me chum é dheanadh* DAR NŪ hǎ mƏ cŪn .ē yiánƏ, *chum é mharbhadh* CON ē vàRŪ.

5. *u* = au under conditions mentioned § 4. *puncamhail* PAUNKŪl from punctum, *suncum, sancum?* sauNKM´, *thá sé baint s. as a(n) gcnámh* hǎ sē Bànt s. as Ə GNǎv, said of a dog enjoying a bone, M. *int ungar 7 int ocras* N´ TAUNGR´, T. L. Eng. 'hunger'. Eng. 'young' pron. īUN is sometimes yauN, 'Jim' yauN = *sēmus óg, stumpa* sTAUMPƏ, Eng. 'stump'.

6. *u* = e in *rud* RED and ROD O. I. *rét.* A shortened form was torn from unaccented position in *aonrét* to *aorred* ERƏD and Ə under second accent to o with dental. Cf. *agam* where Ə to u with a labial. *duit* DOT and DET. Vid. ui.

7. *u* rarely = i. *cumadh : ionadh* P. P. 29. Not a sure instance for *ion-* is often UN- but *cumag : oinig* T. G. 48. where *o* is umlauted to *i*.

u = à in *muna bfuil* MÀRƏ; *muna mbeidh (eadh!)* = O. I. *main bed* shows contraction and loss of *u* on account of the attracted accent MRÀC.

2. The Long Vowels.

á

§ 15, 1. *á* = ǎ, rarely ā. *grádh, áthas* ǎhis, *atá, lán garrán* GRǍN, a grove, *garrán a(n) bhile* the grove of the (remarkable) bush, *garrán in mhuirris* now Garden-a-Morris near *Cill* in the Dēsi, *Garrán fada* now Graunfodda. *cnáb* from *canáb-* the first a lost as in *garrán* before the accent (§ 2, 2, I.) which was attracted by the heavy *b* ending. Hence the lengthening of a. *bráca* a way-side bothy for the fever-stricken, *ádh.*

2. *a* = ǎ often with or through contraction; *bog pál* 'hit the road', from *pábhail*, Eng. paving = pavement, M. Song.

Fål *füghbháil*. Lȧəs *lámhas*, T. L. *láig* 3 per pret. *rámha*
D. R. 98, *gádha* ib. 99. *táirne* = *tabhairne* a 'tavern', *táirne*
= *táraingtle*, O'Neil, *sáthas* sȧəs, T. G. (Des.) *tláthas* TLȧs,
C. M. O. (Thom.) *lághach* Lȧc, *portcladhach* PURTLȧc, now
'Portlaw', *adhamh* åv, T. G. 19, *amhgar* åGR´ with syllablic
division separating *mh* from *g*. Otherwise it should be auGR´.
Cf. *adh-bhar* åvR but *adhbhar* ouR. *Maighistear* : *crádh* T. G. 8,
Dēsi MOISTR´.

3. *a* contracts irregularly in the word *námha* an enemy
owing to the mixing of cases. NōD Carrick-Shock song. (gen.
become nominative) NåD P. Denn. Nūd pl. Nūdə, S. R. NOUD,
Com. song, : *domhan*, P. Denn 86. *novid* : *a gaithib*, ib. 63,
numhid ib. 66 *naoid* ib. 66 pl.? *le naimhde* : *clí-gheal* Sheehy,
dá naimhde : ī *Anna*, : *dithcheall* C. M. O. 16, *namhad* : *áird*
P. P. 142.

4. *a* appears in some words=ā. In the western Dēsi å
in place names and surnames=ā when speaking English.
Cullinane KOlṇāN, *leanán* lṇaN, *cúl dubhán?* KUL DƏvāN. So
in Eng. 'water', 'morning' are wāTR´ MārnṆ. From contraction
baile-atha-cliath BlākIī, *meadhón* māN. *coimeud* O. I. *comét?* is
kimāD. *bán*=a grass field about to be ploughed up, is BåN
when one speaks Irish but BāN as a loan word in Eng. Even
so the place and personal names like Affane, Cullinan, as aíāN,
etc. are used only in speaking Eng. ā being looked upon as
less vulgar and Irish than å. When a further grade of
gentility is reached people so unfortunate as to possess a
name with -*án* auslaut make the last refinement and
change ā to ē, like Eng. *ā* in 'fate'. Then Cullinan
becomes KOlṇēn, etc. In Cass. 143 are found *ansa* ¦niom-
laan, go hiomlaan, le hiomlaan and chiomaad. The last
word still sounded with ā gives a key to the phonetic value
of *aa* and therefore *á* in certain instances had not gone over
to å in the beginning of the present century in Tipperary. O. I.

e is lengthened to ā before *rr* in *fearr* and *gearr*, comp. farə, gir'c.

There appears also a functional use of ā from *á* to embitter the intonation in scornful passages. In a fish assembly convened once upon a time to elect a king the fluke happened to be absent. Hearing afterwards that the appointment had fallen to the herring „*o in sgadán*“ (SGUDĀN) *arsa sé mar sin le searbhas, agus dá chomhartha sin héin thá caime 'na bheul ón lá sin go dtí in lá 'ndiu. Séan de Bhál* Mount Bolton.

é.

§ 16,1. *é* is written only in anslaut as like *e* it always makes a digraph with the timbre-index of a following consonant. *mé* accented *'nar bháil lé mé chur*, D. R. 32 mē, *cré*, kr'c̄, or almost kir'c̄. So *sé* six, *clé, inté .|. inti*, when strongly accented *té; sin preap, té thicfeadh é* sin pr'ap tc̄ hikuc c̄, that is a puzzle ('problem'?) if there was anybody there who could understand it. *ba é* BU yc̄, from *bad e. ceadé* an idle stroller cf. Anglo-Scots 'caddie', *ré* D. R. 94.

2. *é=ī* in *tré* tr'ī, *faoi* usually fé but occasionally fī *fī dhéin*, Com. song. *bodaigh bhoga an phrácáis dá gcárna fí ghleo* fī ſō. Ph. H. *fo thuaidh* is now ō hwuəg.

ī.

§ 17,1. I. ī. I does not appear as representative of a single *i* sign because it occurs only after a broad consonant and so forms a digraph with the broad timbre sign. Thus O. I. *cridě* is now written *croidhe .|.* I+y to I. It is the sound of several digraphs *oi+y ui+y*, etc. and the trigraph *aoi* q. v. It appears as *i* in anlaut after a broad consonant in context. Hence the effort of several Munster writers to represent the pronoun *i* by *ui* after a broad consonant.

2. *i* is written before slender consonants and in anslaut. *min* mīn, *spíce* spīkə a nail, Eng. 'spike', *chidhinn* c̄īiŋ I used to see, *giústis* gūstis, *ní* nī, *lítis* līitis a lily.

An ī arises from *i* in Thom. Des. under the conditions

iu § 7, 1. *cinn* kīŋ, *tinn* tīŋ, Dēsi ai, taiŋ. So from *i* to ī in pronouns and prepositional pronouns where in the Dēsi the radical sound is preserved. *sinn* sīŋ, Dēsi siŋ, *linn* līŋ, Dēsi liŋ. *í* becomes i in *righte* vid. §·2, 3.

ó

§ 18, 1. *ó* = ō is written before broad consonants and in auslaut. Before slender consonants *ói* = ō, for long vowels are not easily umlauted. *bó* bō, *ól* ōʟ, *tóg* tōɢ, *omós* and *fomós, omoas,* Cass. 145. *córda* kōʀdə, *tósdalach* = *toicheastalach* tōsdʟˊuc, *ótis* ōtis, *ar nós na luiche lóg aon oidhche* ər nūs nə ʟᵻhə ʟōɢ en lhə a pregnancy of one night, *spólla* spōʟə D. R. 40, *trócaire* tʀōkərˊə, *lón* ʟōn T. G. 5 should be ʟūn. Vid. § 18, 3. *flós* fʟōs Lat. *flōs* T. G. 87. *trionóid* from Lat. *trinitātem. pionnós* T. G. 3.

2. *ó* from *a* by lengthening in position. Chiefly Des. *sómplaig : criochnóir* T. G. 4. In the Dēsi saump-, *ómbrach:ó* T. G. 23. (Fr. *ambre*) *seomra* sōmrˊə, Dēsi saumrˊə; so *ómhanda : seolta* P. P. 120 *cómhradh : beo táid,* ib. 308.

3. *ó* near *n* took a dark tint and became ū. The following and some other words show this peculiarity. *nó, nóra, stróinse móin* gen. *móna* (pl. *muinteacha meithe S. na Sróna.) fuindeóg spónóg* spūnūɢ Eng. ˈspoon', *tempul fhenóch* a place name, *nós* S. R. *pónaire* pūərə S. R. *bunóc* an infant, *crón, cróna. triúrar* (= *triur fhear) ay crú na cróna triúrar ag tochus a d-tóna agus dheunadh aoinne amháin a(n)gnó* where *cróna, tóna, gnó* are kʀūnə, tūnə, ɢnˊū, *crónán, seóinin* sūnīn one of Eng. descent from *Seán* or *Seón* ˈJohn', *óinseach, ˈnneósad* I will tell, *nóinín, rón* a seal pl. *rónta,* Com. song. *dónti? springiona* great jumps, S. R. *óinmhid* ūnid, Bob, *brón, brónach:cumha, Anna. lonradh : smúit* Sheehy, *aimdheóin* iŋūn, *inneóin* an anvil also iŋūn, *nóment : cúirt* Denn, 86, *Cill mogimóg* a place name, *cnó, trathnóna* tʀăcnūnə, *sgeóin* sgūn.

Without *n; mór,* O. I. *már* and *mór,* Welsh *mawr,* comp.

mó MŪ, *commórtas* comparison, *bórd* BŌRD and BŪRD, lent from Eng. when *ō* = *ū* on account of the labial.

In Thom. Des. not so many words are affected. *mórga : órga:dóirse* T. G. *brónach:deorach* Sheehy. Thom. *móna : bóthar* C. M. O.

lón : trócaire Denn 69, and so spoken, is an exception. *nóta* NŌTƆ is sustained by Eng. ʻnoteʼ. Such instances as *ó'n, órna ró-nádúrtha* are not affected by the nasal.

4. *o* with w from *gh, dh,* etc., in auslaut contracts to ō. *slógh* SLŌ, *clódh* KLŌ, *sógh* SŌ, *sóghach* SŌC, with vocalic opening after w to ou, *rogha, togha,* etc. § 5, 1.

ú

§ 19, 1. *ú* with broad timbre i. e. after a broad consonant = Ū after a slender = ū. *súl* SŪL, but *fiu* fyū, *siubhal* SŪL. An instance of change in the same word is furnished by *úd* ŪD, but *siúd* SŪD.

2. *ú* appears alone before broad consonants and in auslaut. *úr, crú, lúth, cúl, lúdaidhe* a lazy fellow. It represents the Norman-Eng. ou in old loan words, *prisún, gúna, púnt, cúrsa, i gcúrsaidhe* as for, with regard to, *gallún.*

3. *ú* arises often from *u* by contraction. *mumha* MŪ or MŪWƆ, *umhal* ŪL, *ul* Cat. 22. by position of mute and liquid *búnsaig : mórgaig* .|. MŪRGIG C. M. O. 8, *bunsaig* LU. p. 23b, 27. W. from *bun.* Cf. *buinneán* BWIŋǎN, *súgradh* from *su-*SUGR' Dẹsi.

4. *ú* = Ɔ detoned before the accent, *siúd é* hiDĒ.

3. Digraphs and Diphthongs.

§ 20. *ae* = E is historically a middle grade between the O. I. diphthongs *ai oi* and modern *ao.* It is still sometimes written chiefly in auslaut, for a canon of the new orthography forbids *e* to touch any following consonant broad or slender. *rae, lae* gen. of *lá,* so sometimes *aerach* for later *aorach.*

§ 21, 1. $ai = a$ + a slender consonant. Here three possibilities arise a) a preserved, b) gone over to à, c) umlauted to i. There are numerous examples of a) and b). a preserved; *aicíd, ainm* anm, *ait* at, *ais, thar m'ais* har mas, *aitheanta* commandments, *na hahanta*, Cass. 139. *aisge, bean ar meisge bean in aisge, ad = fad(?) 'n ad is bhi sé* n' aD is vi sē, *daingean, airgead, aingcis* anis, *aigne* agnə, *taise, taithniomhoch* taɳuvuc, but *caill* ĸoil, § 8, 4, *aithne* Keatynge, *aithis : leaca* T. G. 32, *aithreachas* T. G. 27, but Dēsi ʀácis for heavy c, §§ 10, 1. and so *aith* fell away before the accent. *aifrionn, saithe* a swarm: *treasgairt* T. G. 57, *aici* akə T. G. 4. Dēsi əkí accented like all compounds of *ag* § 2, 2, V. *aingeal* anʟ', *aibig, sraim* congealed mucus from the eyes D. R. 104, *aisiog, aisdear, aith-, aircluachra* a newt, āɾt-ʟŪcʀ' Dési; *dair*. a is preserved in accented anlaut except in cases of y influence § 6, 1. It stands also after an anlaut dental.

2. *ai* becomes à under the darkening influence of a foregoing labial or guttural (including gutt. ʟ) under the accent. *bain* ʙån, *fairsing* ғårsin, so *flaitheas, maith, caith, caillín, maise, laidin, Cúilín na faithche* in Carrick, *gaise*.

3. *ai* is found umlauted to i in a few instances. *Uaimh na caorach glaise* ɢʟɪsə, the older limestone cave near *Sceichín na raince* in Tipperary, from *glass, maille : duille* T. G. 12, *glaine* ɢʟɪnə, a glass, from *glan, clainne* ĸʟɪɳe ·from *cland, cait* ĸɪt, gen. of *cat*.

4. *ai* under conditions mentioned §§ 7. 8,4. 6,2. to ai, oi, î. *crainn* ĸʀain, *raince* ʀainkə, *frainc* ғʀaink, *caint* ĸoint, *maighistear* moistʀ' *baintreabhach* ʙointɾuc, *laigheann* ʟiɴ, *saidhbhreas* sîvris *maidhm* ᴍîm, *saighnean* sînåɴ, (with old *én* to åɴ as usual) *saighdiúr* sîdŪʀ or sɪʙŪʀ, *aighneas* înis, *saidhbhir : gadhair*, in *pe 'cu bocht nó saidhbir sinn ní cuideachta do gadhair(ibh)sinn*, My Father. *aibhne* inə rivers.

An ē sound appears to take the place of this î sometimes in Thom. *saighid : sgémh* P. P. 90, *an Mangaire Súgach,*

Limerick. *dha saod triom : péin*, O'Neil. Also = ī, Des. *saoigheada : tríomsa*, *M. ní Dhonagán, taidhbhreadh : trid* D. R. 66, Dēsi ᴛîᴠʀᴜɢ, *aighnis : díth*, D. R. 88.

5. *ai* = e which is an umlaut of *o. raibh* ʀᴇᴠ = *robhí* and with *ro* accented after *ní* to *roibh* and with *i* umlaut of *o* to ʀᴇᴠ, ʀᴇ before pronouns. *saidhbhir* sîr´ or sᴇir´ and sevr̥´ an index of its etymology **so-id-ber*. O. I. preps. *for* and *ar* fell together (like *la* and *fri* and, with mistake of *f* for prothetic, through *ri* to *le)* The forms are; *orm, ort, air, airthe, orainn, oraibh, ortha.* The masculine pronominal form *air* is used also for the simple prep. as *aige* for *ag.* The pronunciation of all the forms goes back to *for*, with loss of *f* through confusion with prothetic *f*, the writing of *air, airthe* comes apparently from *ar.* The pronunciation is however er, erhə, *ai* being merely the phonetic for umlauted *o.*

6. *ai* unaccented = ə. *cabhair* ᴋᴏᴜr̥´, *tabhair* ᴛᴏᴜr̥´, *Corcaig* ᴋᴏʀᴋig, *caisleán* ᴋɪsᴌẚɴ, contraction *gabhail leo* ɢōl, C. M. O. 10. Before the accent *ag aireachas* i gr´ẚcis.

7. *ai* = o in *ghaibh sē* ʏᴏ sē̃.

8. *ai* = ẚ in *aithrighe* ẚrhī, where rh constitutes position. In this combination the *r* is unvoiced a fact to which Dr Pedersen directed my attention.

§ 22,1. *ái=*ẚ with a slender consonant following. *táinic* h̃ẚnig, *táinic sé* h̃ẚnə sē̃, *páirc* pẚrk Eng. ʻpark'.

  2. *ái* in *paráiste=*ō. ᴘʀōstə, (from French *oe*), *paroisde* Cat. 30.

§ 23,1. *ao=*ᴇ. It occurs in anlaut and inlaut; *ae* is reserved for a few auslaut occurrences. *ar aon chor* ər ᴇᴄʀ´, *taobh* ᴛᴇᴠ, *gaol* ɢᴇʟ, *saogal* sᴇʟ, sometimes sᴇɢʟ´ in verse T. G. 16,23, either through remembrance of the unaffected consonant or (more likely) from the traditional word-picture. Those Latin loan words exhibiting a media for tenuis i. e. Welsh vocalic infection, *saoghal*, etc. came to us through a British medium. In genitives vocatives and plurals of *o* stems

E is umlauted to I by the following slender consonant. Both E and Ī are broad guttural vowels, E being broad throughout, I only at the beginning as it tapers to slender at the end. Hence the former stands between broad consonants the latter between a broad and a slender. Western usage does not make this distinction and perhaps goes back to the umlaut form. After a labial=WE, *faobhar* FWER. The word *laogh* a calf makes pl. LIg, where g is an instance of Dēsi auslaut g fondness.

2. *ao* in *caora*=Ī KĪRə, gen. however KERUC.

§ 24,1. *aoi* is a trigraph the technical phonetic sign for I. It arises from O. I. *ai, oi*, slender, or is the umlaut of *ao* before a slender consonant. It may also come from ī in sentence sandhi after a broad consonant. Wherever possible in the Dēsi the timbres of auslaut and inlaut vowels and consonants are accommodated to each other. Cf. *aoin neach* aiŋə, from *aon neach, an bhean sin* N´ vaN sN´, but *na bainbhidhe sin* Nə Bǎnī sin (with loss of v before *-idhe* of new pl.). So *iad súd* but *i siúd*. Hence the writing of *i* the pron. of 3 per. sing. fem. as *uí* after a broad consonant. *baois* BwĪs *go maoilín an tsléibhe* GU MwĪlīn ņ tlē. *aoi* stands for *-adh* in accented auslaut in the word *claoi* KLĪ, from *claidhe*, pl. KLahUCə, *dlaoi* (?) a lock of hair, *dlathach* adj.

2. *aoi*=ē in *faoi* O. I. *fo*. Sometimes fī in ˌsongs. Com. Killown.

3. *aoi*=ai before y of *nn* in *aoin neach* aiŋə though the combination is open.

4. *aoi* is used as an orthographic device to express I the broad-slender vowel in a difficulty arising from a merciless use of the *caol-leathan* rule, *dreachtaoin* for *drechtín*, *cht* always resists palatalization, *atámaoid* for *atáimid*. O. I. *ataam*.

5. The group *au* does not occur. Older *au* in such words as *auctardhas, audhacht* is now written *u. ughdarthás, uadhacht.* *avios, cf. Lat. *avos*, became *aue* and finally the

O of surnames. In gen. and pl. it is umlauted to I written *i* and *ui* from *avī, now however ī or I according to the foregoing timbre. *árd* makes derivatives *aoirde* and *áirde*. *a h-aoirde cheart*, C. M. O. *is barr a dhá chluais anáirde*. *Maidrin Ruadh* song.

ea

§ 25,1. *ea* = a. This a is the broad umlaut of O. I. slender-broad *è* itself of various origins. Thus *ben* from *benā, *fer* from *firos. The influence of the broad consonant working farther opened *e* to *a* slender-broad. *seanbhean* saNə vaN, *seanduine* saNDINə, but *seandrui* saᴜNDRĪ and *seanriabhach* saᴜNRĪᴜc, the old grey cow's days, the cold weather at the beginning of April. *lean* laN, *leath* lac, *leathcheann* la-caᴜN, a head bent to one side, half the head, in perpendicular section, *ceatha* kahə (gen. of *cioth* kᴜc,) *bean feassa* baN fasə, *greannmhar* gr⁏aNᴜʀ, *'seadh* sa.

2. *ea* before labials and gutturals (including guttural L) = à. *eagla* àGL⁏ə, *preabadh* pr⁏aʙə, *eaglais* : *carthanach*, T. G. 5. (but *freagair* ꞮꞮraGꞮ⁏ and *dealbh* daliv) *Cluain gheal Meala* KLᴜən yàl màlə (for l in *meala*, *bealach* etc. vid. Pedersen, *Aspirationen i Irsk.)* *mo chreach* ᴍᴜ cr⁏àc, *Sighle na gcearc* sīlə Nə gàʀK, *seacht* sàcᴛ, *do breabaigh* : *ganguid* Sheehy. *dhearcas* yàʀKis, *leaba* làʙe, gen. làᴘņ, *gealaighe* gàʟĪ, (gen. of *gealach* glàc), *sceach* sgàc, *ceapaire* kàᴘiꞋʹə, *neamh* nyàv, *leamh* l(y)àv, *seach* sàc, *isleach* istàc, *seachaint*, sàcņt, *sgeamhaighil* sgàvīl howling of dogs, *treabh* tr⁏àv. Here the dark timbre consonants radiated an umlaut into the word.

3. *ea* = au under accent before *mh* and the heavy combinations inducing it. § 4,1. Unaccented = ᴜ, before accent = ə or nul. Thom. Des. accented = ou, ᴜ. *feall* fyaᴜʟ, *peann* paᴜN. A heavy liquid combination resolved does not induce anusvâra and the original sound is preserved, *peanna-chinn na cruinne* paNə caiɲ, *deallradh*, daᴜʀə, daᴜʀ, *sgannradh* sGaᴜʀə, sGaᴜʀ *(lr, nr*, in Arran, Pedersen, 23),

greanda : cubhartha Thom. *reamhar* ʀᴀᴜʀ, ʀᴏᴜʀ, M. Des. song. *mealltach* ᴍᴀᴜʟᴛᴜᴄ, *mh* remains in position. *dream* draᴜᴍ, *leamhnacht* lauɴᴜᴄᴛ from *leamh-lacht* = luke-warm milk, *teampul* tauᴍᴘʟ′ or ᴛᴀᴜᴍᴘʟ′, *cleamhnas* klauɴɪs *simróg* in Dunne for *seamróg* where *ó* drew the accent. *Teamhair na Rig* tȧvr̥, Sheehy, *geancach* gauɴᴋᴜᴄ. Also the Eng. *young* is often yauɴ.

4. *ea* in *beag* = e. This is against Dr Pedersen's statement that „ea paa Irsk ellers uden Undtagelse udtales som a; netop Udtalen med e breviser," p. 27. In a conversation with me he derived *beag* from the old dat. *biucc* used adverbially. In the western pronunciation we have no doubt the old dat. and uses of old dat. or acc. for nom. are plentiful in all modern Irish, but *iu* cannot have been the forerunner of *e* in the Dēsi pronunciation. Besides the English represent the *-beag* element of place names by *-beg* and so they must have heard it pronounced when they first came amongst us. It is possible that the numerous occurrences of such place names pronounced -beg in Eng. may have sustained the old *e* in Irish. The historical Eng. rendering of Irish place-names being the product of an independent tradition is of some worth in dealing with Irish sounds and occasionally preserves an etymology disguised in the native version.

5. *ea* = ou in contraction with w from *bh* in inlaut § 5,1. *threabhfainn* hrouhin̦, Bob, *feabhas* fouis, *seabhac* souᴋ, a hawk. This is a Germanic word borrowed by the Welsh from the Saxons and ʻtranslatedʼ into Irish by the artificial substitution of *s* for *h*, an analogy from the knowledge that Irish *s* was the equivalent of Welsh *h* in Keltic words. Cf. the same change between *c* and *p* in Irish *cáisc* Welsh-Latin *pasc*, etc. *creabhar* kr′ouʀ′ a wood-cock and the horse-fly called in Anglo-Irish a blîn dȧᴄᴛʀ′ .|. blind doctor.

Unaccented *duilleabhar* ᴅʟūʀ′.

6. *ea* with y of *mh, gh, ch* after a liquid in auslaut syllable = ī. *inghean* inīɴ, *coitcheann* ᴋᴏᴛīɴ § 11,4.

38

7. *ea* is sometimes written for *eu* = O. I. *ē. leadmhar*
C. M. O. 23, 'daring' from *lámhaim.*

8. *ea* unaccented = ə or nul. *intleacht* întlucт, *gealach*
glạc, *simróg* Dunne, *glé-geal* glēgʟ', with second *g* unaffected.

9. *ea* representing O. I. *e* = a is lengthened to ā, å
Thom. Des. before *r* in position or in contraction with w
from *dh, gh. ceardcha* kārтə a forge gen. kārтṇ, *meadhchaint*
from *med* wage, lanx, W., мācnt, Thom. мåcṇt, *agus ualach
sgadán am mheadhchaint ar thaobh de : sgadán,* D. R. 44.
spleadhchas sрlåcis from *spled gone to å even in the Dēsi
neamhspleadhach na-sрlåc, *neamhdha : áluinn* T. G. 22, nåfə,
cneadh : bláth T. G. 7, pl. knåhə : *cráidhte, Zeit. f. Kelt. Phil.*
I. p. 142, : *páis,* Denn, *cneadhaire* knåər'ə, T. G. 34, *breaghdha*
br'åhə, now comparative. (Vid. Manners and Customs Vol. III
p. 183, *bregda* .|. *bricin* 'threads of various colours'; the same
word also occurs in an old poem quoted by O'Curry), *na dheadh
sin* often in poetry nə yå sn' 'after that' O. I. *degaid, leaghadh*
lyå to melt, *sneadh* snå, nits. So the names *o Seagha, o Deagha.*
So *ei* in *leigheamh* lēəv, to read.

Examples of lengthening from *r* in position like *ceardcha*
above. *ceathrmhadh* kārhu, *dearna* dåʀnə palm of the hand,
and form of *deanamh, dearbh (?)* dāʀuv, appearance=*deallradh,
ferr gerr,* fāʀ gāʀ, *téarmadh : eadach,* T. G. 10, *ceafrach : cárnach,
M. Súgach* P. P. 90. 'capering', *do cheáthfrainn* T. G. 26,
Seaghan sån 'John'.

10. *ea*=î with y from *dh gh* in position (assimilated
d, g? Cf. *snaidm* snîm, pl. snåмn'ə as if from *snam-manna)
meidhg, meadg,* mîɢ, *treaghdadh* tr'Ɪᴅə, *Tadg, fadb, teaghlach,* etc.

11. *ea*=ū by analogy, *cuigeadh* kūgū, *cuigu,* Cat. 27,
seiseadh sēgū, *seishu,* Cat. 27.

§ 26,1. *ei*=e+slender consonant. Old *e* usually preserved
its sound before a slender consonant. Vid. *ai.*

leis les, *sgeiche* sgehə, *beir* ber, *te* te, there being no
necessity for writing the digraph in auslaut. Comparative *teo,*

also pl., O. I. *téit* being lost. *leinbh* leniv but also umlauted
liniv, *beidhead* bec seems a mixture of subj. 3 *beith* and conj.
biad. To join the vowel sound of *beith* to the broad consonant
of *biad* it was thought necessary to add on -*eadh, muna
mbeidheadh* is pronounced mrˈàc, but màrə vec ʻonly forʼ
M. song.

2. *ei* umlauted to i. *teine* tine, *leinbh* liniv, *deitheanas*
from *déineas* and *th*=h developed and caused shortening as
usual, dihənis, *leigint* ligint, *reilg* rilig:*cinedh, M. ni Dhonagán.
deithbhir* difɍ, *eidir* idrˈ and dir, *dar* a *bhfcillˈbheart* dar ə vilə-
vart, *gein, gin* Cat. 8, *meithil* mihḷ, *neimh* niv, *sein* in *sein-
shean-athair* sinhaɴahirˈ great-grandfather, *meic* mik.

3. *ei* in groups with y=ei. This in the variant from î
caused by the e element § 8,1. Sometimes the sound appro-
ached ē. *eirghe: céile* T. G. 11, sometimes again not distinguished from î. *meidhir* mɛirˈ *bhfeigil* vɛil. *r* too seems to
have an effect, hence Eng. ʻpairʼ ʻstairsʼ are pronounced pɛirˈə,
stɛirˈə, and often written *perdhire staghaire;* for as y from
dh, gh, gives this sound so those consonants are written as
phonetic signs of its presence. *dh* is however sometimes used
as a mere syllable divider or glide as in the case of *mi(odh)údh.*

éi

§ 27, 1. *éi* = ē with slender following consonant, *eu* = ē
with a broad consonant following. *pléidh* plead, go to law,
ag pléidh leis squabbling, *bréig* dat. of *breug, Éire, na dhéig sin*
Denn 69, *tóirtéiscach (?)* of prococious manners (of a lad, bad
sense), *gréithre* grēhrˈə wares, chattels, *a réir* last night, *araoir*
P. P. 175. Vid. Stokes in K. Z. XXXV, p. 592.

2. There seems a lengthening before -*st* in *éist* Thom.
Dēsi ɛist, *pléist.* Before *r* in position *péirse* Eng. ʻperchʼ.
So Eng. ʻearns,ʼ ʻfernsʼ is ērnz, fērnz.

eo

§ 28, 1. *eu* is mostly long. It is short = o in ocrˈ, a
key, a border, *eochair-aoibhne : codlaim* Keatynge, *Eochaidh*

ocī, *seoch* soc, *seochas* socis, 'compared with,' translated 'towards' by bi-linguists.

2. *eo* long = ō. *leo* lō, *eolus* ōʟis, *deor* dōr´ (from *dēr.* Cf. *scēl* sceol, *bēl beoil),* *teo* tō, *na deoigh* dat. of *diad* ɴə dōəg, *ceol* kōl, *leoghan* lōeɴ, *leomhaidh* redup. fut. of *lamhaim* and with *f* .|. *v* + *h* to lōꜰuc, cond., *geocach* gōкuc D. R. 87.

3. *eo* after labials = yō. *beo* byō, *ar imeal do bheoil a stóir mo chroidhe se* vyōl, Sheehy, *meoin* obl. case of *mian* and used as alternate nom. for rhyme, *feorach* fyōʀuc gen. of *indfeoir* the „Nore!" *feoil, feola* gen. fy-, *feogh* fyō interj. of displeasure at a disagreeable smell. Here it is doubtful whether the value of y should be given to the glide. Neither does the rule hold universally; many labials have no following y. *meoin* is doubtful as it is no longer heard.

§ 29. *eoi* is a trigraph = *eo* before a slender consonant. The sound is ō. *sgeoig* sgōəg, *ceoil* kōl´, *sgeoin* a fright sgūɴ ō to ꞟ by reason of the nasal and ꞟ = ū after a slender consonant.

eu

§ 30, 1. *eu* (= O. I. *ē* + a broad consonant) is a new diphthong = ī´a, īá; occasionally the old sound of *ē* is also preserved. *meur* mī´aʀ, *feur* fī´aʀ, (gen. *féir*, fēr), *deunadh* dīáɴə but *dēnuv* Denn 65. *leus* lī´as, *ni'n leus céille aici* nīn lī´as kēl əki, *eun* ī´aɴ, but *ceol na n-eun* ēɴ, Com. song. *leusadh* lēsə, Bob, *eug,* īaɢ but *gairid go n-eugfadh* ɢu nēкuc. *treun* ē and ī´a, *deug* dīáɢ, *breun* br´īaɴ, *bheursa* vērsə, perhaps supported by Eng. 'verse,' *ceudna* kīáɴə, *peucach* *piacach,* Denn 65, *ceud* kiáᴅ, a being strongly accented, ī is only half long. So *ceudna. meud* mēᴅ, old *ē* preserved even before dental ᴅ, *teagair* tīaɢɼ´.

From lengthening through loss of *n; eudóchas* iaᴅō´cis, ō draws the accent and both i and a are only half long, *eudtrom* iáᴅʀ´ᴍ, *eugcóir* īaɢōɼ´.

2. *eu* before labials and gutturals including guttural
L = īà (yà). *muineal* MUNĪàL, *sgeul* sgīàl, *feuch* fīàc, (Des.
fech, Cat. 8), *feuch air* fyàc ŗ', *eugmais* (= *ēmmais)* iàmis,
eudmonn iàMN' and yàMN'.

ia

§ 31, 1. *ia* in O. I. is a diphthong = īə. *Dia* dīə, *īasg*
īəsG, *grian* gr'īəN, *do chiap* cīəP, Də hcP(?) S. R. *iarfaidhe*
īrhī, by loss of accent to -*aidhe, cad iad* KàD īəD, *cad eud*
Cass. 139 = *catēt* .|. *cadéd, mias* mīəs, *biadh* (now *bidheadh*
where *dh* is used to divide a diphthong made dissyllabic.)
iarla īəRLə from Skandinavian *iarlr*. A consonant relic is
silent after *ia. liagh* līə, *cliamhan* klīəN or TlīəN, *diabhal*
dīəL, *bliadhain* blīcn, *fiadh* fīə, *fiadhta* fīcTə. *riaghail*: *stiúir*
T. G. 2, = Eng. 'rule', ordinarily rīəL, or is at most a glide.

2. *dia* .|. day in names of week days is always pro-
nounced dē, the old locative become nom.

io

§ 32, 1. *io* short and long. Short = i with broad following
consonant. Its occurrence in final noun syllables usually
denotes *u* stems. *fios* (O. I. *fiss)* fis, *lios* lis, *slios* slis, *sprios*
spr'is *bior* biR, *spiorad* sPRiD by metathesis. *siosarnach*
sisR'NUC, *iothalla* hagyard, ihULə, but ahULə Com. song. *cion*
chin, Cat. 26, kuN Dēsi, vid. infra. *Sionnaine:* whiggiona
P. P. 96, *iongantas: mithid* C. M. O. 30, ŪNTis and exceptionally auNTis Dēsi.

2. *iodh* short accented before *bh, l,* etc. § 6, 1. = î;
long = ī in the same situation. *tiodhlaicthe* tilikəhə, *iodhna* îNə,
pangs, but long *biodhbha* bīvə, *iodhbairt* ībirt, *ibirt*, Cat. 22.

3. *io* before gutturals and labials = v. *liom* luM, lauM
exceptionally in the song *Páistín fionn, sliocht* sluCT, *crioth*
kr'uc: *cnoc* P. P. 248, *cioth* kuc, gen. kahə, *tiocfaidh* TÚKig,
with pron. TUKə sē, *cionnus* KUNis, *cunas*, Cat. 26. *cion* kuN
affection, *sioc* suK gen. *gob an t-seaca* GOB ə TΛKə, *biolar, fiolar*
byULR', fyULR', i after labial to y; *humlan* = *h-iomshlán* Cat. 24.

umercach Cat. = *iomarcrach, iomarcad : fulaing* C. M. O. 13. *giolla : tusa* D. R. 56, *tiobaiste* tuʙistə, *iomadach : cumasach* C. M. O. 30. *siolla : cuireadh* D. R. *Eactra an Amaráin.* So *liobursach* untidy, *giolca* ɢyuʟᴋə a place name, but *giolcach* ɢɪʟᴋâc by loss of accent. (§ 2,2, III.), *sprioc, miochair, briocht, riocht, rioch* ʀᴜc, *sgiob* to snatch, *sgiolpadh, triocha*.|. 30.

4. *io* unaccented=ə. *taithniomath* taɳuvᴜc. So it often changes with *ea* in unaccented position. *iomarcradh* when accented on second syllable=əᴍᴜʀᴋə, *thá an iomarcradh cainte agat* hâ ɴ' əᴍᴜʀᴋə ᴋoint əɢᴜᴛ, *timchioll* hoimᴘ|.

5. *io* before the accent often falls away; *iomdha* ᴍᴜ̄, *iona* ɴâ.

6. *io* under the conditions in § 4,1=au. *gan piomp leamh* ᴘᴀᴜᴍᴘ „could" pride (in Sheehy : ī) *lionruith* lauʀᴜc : *isbirt* Sheehy, *tionsgain* ᴛᴀᴜɴsɢɳ, *iomchur* auᴍᴘʀ' from *imbchor* and *b* provected to *p* by aspirate as usual : *múinte* Denn, 61, *ionryic oonric*.|. ᴜ̄ɴʀik Denn. *Fionn* fyauɴ also fauɴ, *mionn*, mauɴ pl. ɴiɴī, *d'iompaig* yauᴍᴘɪɢ with pron. as usual yauᴍᴘə sᴄ̄, *ionta* auɴᴛə and iɴᴛə *ioncholnughadh, aoncholnugh*, Cass. 139, *pionta* ᴘᴀᴜɴᴛə, Eng. 'pint' *prionsa* pr'auɴsə 'prince', sometimes *prúnsa* in songs, *iontaobh, auntee* in Dunne's phonetic rendering of the Carrick-Shock song.

The prep. *ind-* as noun prefix shows the same irregularity of intonation as *con.* ˙ In some instances even though accented it must not have grown on to its word and the heavy combination necessary to induce the slurred diphthong au was not formed. Cf. the different treatment of *seanduine* saɴᴅɪɴə and *seandrui* sauɴᴅʀī. Hence in a good many instances we find the unaccented intonation of *ind; iongantas* ᴜ̄ɴᴛis, but auɴᴛis, Tom Lannon in a recital.

§ 33. *io* lang=ī with a broad consonant. The *o* element was chosen as timbre index for *a* would have induced confusion with the diphthong *ia. miol* mīl, *mio-*=*mi-*+a broad consonant, *miorbhuil* mīʀᴜ̄| and miʀᴜ̄| from Lat. *mirabile* with contraction and vocalization of *v* after a liquid in auslaut syllable § 10,7. So *nior*, *píob*, *riogdha* rīgə, *síol*, *síor*, *sios*, *cionádh* T. G. 90. As after the long diphthong *ia* so here a consonant relic is absorbed. *biodhgadh* bīgə *biodhbha : fior-lag* Sheehy, *fioghair*, *fior*, Cass. 139, *Cliodhna* klīx̊å S. R.

iu.

§ 34. *iu* short and long. Short=ʊ with foregoing slender consonant. *fiuchadh* fyʊcə ebullition, *bord an sgiulpa (sgiolptha?)* sgʊlpə 'snatch-mess', *sgiub* sgʊʙ pluck, carry off, *piuc* pyʊk a morsel, *tiubaist* tʊʙist, *indiu* əɲʊv, with narrowing of lip-rounding until a spirans was produced, *niuv* Cat. 10.

§ 35. *iu* long, an old diphthong, now=ū, yū after a labial. *ciumhais* kūis, *fiu* fyū, *siúr* sūʀ, T. G. 71, *paisiúnta* pasū̃ɴtə from Norman *passioun*, *siúcre* sūkṛə *diugadh* dūgə, *siubhal* sūəʟ, or sūʟ.

§ 36. *iui*=iū with a slender following consonant. *liuireach* l'ūr'ʊc T. G. 6, *Siúir* sūr' the river 'Suir', *miliúin* T. G. 23, *tiúin* ib. 43, *puisiúinidhe* pwisūni a kitten, *giuis* gūs, *ciúin* kūn.

oi.

§ 37,1. *oi* short and long. Short=o with slender following consonant. *foithin* ꜰohɲ, *cois* ʀos, *coiscéim* ʀisgēm with *c* preserved but to *g* because *sc=sg* except in auslaut. *ni iarfainn ach naoi g-coiscéim chum dul de léim tar geata,* ni yīʀhiɲ âc ɴɛ ɢisgēm cɴ' ᴅoʟ də lēm haʀ gatə *Seaghan O Briste leathair* song, *coisgim.* So *doi-=do-* before a slender consonant. *doirtim* ᴅoʀtɱ. Also *foi-, fo-, foirbthe* ꜰoʀəfe, *roith* ʀoh, *sgoith* sgoh. The aspirate quality of the *th* in such auslauts announces itself by the explosive character of the vowel when in pausa or by the appearance of the aspirate before a vowel. *a bheith h-acharnach,* Bob, *go raibh maith agat* ɢʊ ʀɛv ᴍåhə ɢʊт. *soitheach* sohʊc, (but *sathach* Denn, 61, like *innsa* for *innso*), *coitcheann* ʀotīəɴ, *loit,* T. G. 7. *doilbheas : dorcadh* ib. 6.

2. *oi* umlauted=i. In this condition it often interchanges

with umlauted *ui*. Short vowels succumbed to umlaut influences much easier than long vowels, the palatalization which spread backwards from one element of a compound to another often succeeded in changing the timbre of vowel as well as consonant. Long vowels however were endowed with greater resisting powers.

croise ᴋʀɪsᴇ : *cineadh*, T. G. 7, *coinne* ᴋɪɴə, *roimh* ʀɪᴠ, as prep. pron. ʀoig, *goineadh, ginag*, Denn 80, *goid, an máilin a goideadh uaim* ɢɪᴅᴜɢ, *thoir* hir, *toinnibh* ᴛɪɲiv, Bob. *stoirm: Muire* T. G. 3, *coire* ᴋɪrˈə, ib. 19, *illidh-phiasd=oill-phiast* Ren. 69, p. 34. *soille : truime*, D. R. 90, *coirp* ᴋɪrp, gen. of *corp*, *cnoic* ᴋɴɪᴋ gen. of *cnoc, troighthe* ᴛʀɪhə pl. of *troigh : dlaoi, soip* sɪp gen. of *sop. oiread* irəᴅ and ᴜʀəᴅ, *d'oirfeadh* ʏɪrhᴜᴄ, *oide* idə, and so *oineach, oileán* (and îlᴀɴ) *oireamh, roilg, loise, loinnge, foirseadh, coindeal* ᴋɪɴʟˈ, from *candéla* to *cáindël* and *a* umlauted to *i* by the slender group *nn*. This should be coiɴʟˈ but that the ending is lightened by *l* sonans. (Cf. *coindleoir* ᴋoiɲlōrˈ), *coin* ᴋɪn, pl. of *cú, coille* ᴋɪlə gen. of *coill* ᴋîl, a wood. *coimeud* kimāᴅ. *croidhthe : buille* T. G. 18, *dh, gh*, absorbed before *-the. thoirmeasgann hirmisgan*, Cat. 25.

3. *oi* with y from *dh, gh*, contracts to ī, *croidhe* where *o* is now only the timbre index of the broad group, *troigh : dlaoi, oidhche* ɪhə, *coidche* ᴄɪhə. So with y from *mh, coimhdeacht : oidhche,* T. G. 17.

4. *oi* in *roimhe* 'before him' = oi, ʀoig, also *roimhe sin* before that. ʀɪᴠᴇ sin in Connaught, *roime seo : díbriog* P. P. 308, *roimpe* ʀoimpə, ʀēmpə C. M. O. : *taréis*.

5. *oi* with y from *bhˈ, ghˈ, dh, dhˈ, mˈ*, or developed by *ll* in pausa or *l* in position = i. vid. § 6, 1. *oighre* irˈə, *droighnidhe* ᴅʀɪnī but ᴅʀɪnī Bob. *foighne* ꜰɪɲə, patience, *doimhin* ᴅin, *doighir* ᴅirˈ, *coibhneas?* : ī Sheehy, *soillse* sîlsə, *púca poill* ᴘŪᴋə ᴘîl, a dried toad-stool, *coill* ᴋîl, *coimpleax* has not umlauted *o* nor made its group *-mpl-* slender so it is pronounced ᴋoᴜᴍᴘlīax appetite; in *trompeud* also no umlaut

and so TROUMPĒD, *gan mhoill*, vîl.: î, C. M. O. 1, *thoill*: ī
T. G. 6. *coigthigheas* ᴋîkīs. Unaccented *poillín*: *coinín* ᴘilīn
or ᴘwilīn, D. R. 66.

6. *oi* with *i* umlaut to e in some words. *oiffig* (officium)
efig, *oibre* ebʀə, *soirbiughaàh* serəvu a Western word, *toil* ᴛel,
dat. for nom. (This words shows progress of the umlaut as
three forms are in use. *tol, má sé do thol é* hoʟ, old nom.
without umlaut, and hel and hil), *croiceann*: *loisgithe*, Denn 79,
present Dēsi ᴋʀekṇ, Connaught krek'ṇ, krœk'ṇ Finck, *toice*
ᴛokə and ᴛekə, *Cloichín an mhargadh* a place-name ᴋʟohīn aud
klehīn, with anlaut consonant group also umlauted. *soidear*
sedʀ' T. L. = *sodair*, *oile* now *eile*, *troidfeam* T. G. 44 ᴛʀetм',
but *troid*: *gol* D. R. 68. *h-oileadh*: *leire*, C. M. O. but
oileamuint, *iliunt*, Cat. 16.

7. *oi* with y from in *mh* is sometimes = ē in Des.
coimhtheach: *breugach*: *taodach*, also *coimhtheach*: *geugach*.

8. *oi* unaccented = ə or nul before the accent. *fuirionn*,
foireann, ꜰwiʀ'ɴ', *croicionn*, *croiceann* ᴋʀekɴ', *coilleán* (where *eá*
represents the change from old *ēn* to *án* both vowels being
written) ᴋʝ'åɴ and ᴋl'åɴ, *an dá oiread* ɴ' ᴅå ʀoᴅ, *foiréigion*
ꜰʀēgṇ D. R. 38.

§ 38. *ói* = ō with slender consonaut following. *cóir*
ᴋōʀ', *móide* ᴍōdə, but *mó* ᴍᴜ̄, *mór* ᴍᴜ̄ʀ, *óinmhid* ᴜ̄nid, Bob.
Vid. § 18,3, *cóisir*, *dóirse*, *lóitne* ʟōhnə, with unvoiced *n*,
a breeze, *piolóir* T. G. 3, *Brian Boróimhe*: *slógh* D. R. 36.

ua

39,1. *ua* is a diphthong = ᴜe. *uasal* ᴜisʟ', *suas* sᴜ̄is,
sūis, *cuach* ᴋᴜ̄uc, *buachaill* is ʙᴜ̄ecʝ, *luag* ʟᴜ̄àɢ.

2. *ua* absorbs consonant remains, *uabhar* ᴜəʀ, *uadhacht*
ᴜ̄uᴄᴛ, : *dual*, Denn, 83. So *tuagh*, *ruadh*, *rómar* is ʀᴜ̄ᴍʀ',
buadhartha ʙᴜ̄ʀ'hə, ʙᴜ̄eʀhə.

§ 40,1. *uai* = *ua* before slender consonant = ᴜ̄we or
we. *cluain* ᴋʟᴜ̄wen, *an uair* ɴᴜ̄wer' ɴer' and er' from con-

fusion with *iar n-*, *chuaidh* cūə and hwūə, *uainn* wᴜiŋ weŋ by change of accent, *uaidh* wūig and weg, *uaim* wᴜim and wem, *fuaim* ғūim without w after a labial, *uaigneach* ūəgnᴜc or almost wegnᴜc.

2. *uai* = ᴵ in *smuainte* sᴍīnte, *smynte*, Cat. 9.

ui

§ 41,1. *ui* short and long. Short = ᴜ, o, with slender consonant. *duit* ᴅot also ᴅet (as *rud* gives ʀoᴅ and ʀeᴅ O. I. *rét*), *cuid* ᴋᴜᴅ and with umlaut ᴋɪᴅ, *cuisle* ᴋᴜslə and ᴋɪsle, *amuigh* əmᴜ, *puinn* (Fr. point) ᴘūn from the literature? It should be ᴘaiŋ, *puinn : dlaoi* P. P. 29.

2. *ui* = ᴇ in *buidheachus* ʙᴇcis, seems imported from Des. Cf. *coimhtheach : geugach*.

3. *ui* is mostly umlauted = ɪ. *cuirfead sa* ᴋɪr'həᴅ sə, unvoiced r', *suip*, *soip* sɪᴘ, *cnuic* ᴋɴɪᴋ. So *cluig*, *thuit a lluigire lag aici* hit ə ʟɪgr'ə ʟȧɢ əki, she almost fainted, *cluigchill* ᴋʟɪᴋḷ (*gch* to k) a bell (round) tower, *druimfhionn* ᴅʀɪᴍɴ', *druim* should be ᴅʀîm in pausa, however *drom* is the Dēsi word. So *spuir* 'spurs', *luisne*, *cluithche*, *cuilm:gliocas* T. G. 48. *duibh* ᴅɪᴠ, T. G. 47, *bruingeal* ʙʀɪŋʟ' *muir* ᴍwɪr' *Muire* ᴍwɪr'ə, *ná fuil*, *uile*, *mór-uilc* ᴍūʀɪlk, *cuisne*, *cuithe*, *cuir*, *buile*, *buinneán* ʙwɪŋȧɴ, *buime : milis* T. G. 18, *uireasbach* ɪʀisʙᴜc.

4. *ui* under conditions in § 6,1 to î, with nasal umlaut ai. *tuillte* ᴛîltə, *saidhbhreas : * î, *Anna*, (*muintear*, *suim.* Vid. § 8,4.) *cruinn* ᴋʀaiŋ, *bruinn* ʙʀaiŋ, *cuing* ᴋaiŋ. *cuimhin* ᴋain *: crích* C. M. O. § 7,1.

5. *ui* with umlaut and contraction with y=ī. *luibheanna* C. M. O. 2, ʟīn'ə, *suideamh* sīᴜv, *buidhean* ʙwīɴ', *fuidheach* ғīᴜc and so *muidheamh* ᴍīᴜv, *cuibhreach*, *o Duibhir*, *leabar na h-uidhri* ɴə hīr'ə; so *duidhe*, *duibhe*, *suidhe*, *luighe*, *guidhe* ɢī.

6. *ui* unaccented=ə or nul before the accent. *duilleabhar* ᴅlūʀ, *Uilliam* līᴍ, *fuireach* ғʀȧc, *buinneach* ʙwiŋȧc.

§ 42. *úi*=Ū with slender consonant. *túirseach : dubhach*
Sheehy, ᴛᴜʀѕâc in the Dēsi, lengthened in Des. by *r* in position.
dúil ᴅŪl, so *gnúis, cúis, drúis, cúig* O. I. *cóig, búithreach*
ʙŪrhᴜc T. G. 6, *na súirt* pl. of 'sort' T. G. 32, on analogy
of *órd úird.*

4. The svarabhakti Vowel.

§ 43,1. From what for the present purpose is called the
svar. vowel there must be distinguished the vocal element of
sonant liquids, an incident not treated of here beyond the
marking of occurrences in phonetic script as they arise.

The svar. vowel seems a metathesis of the liquid sonants
ḷ, ṇ, ṛ before labials and gutturals. There are no examples of
occurrence before guttural ʟ as this assimilates with *l, n, r*.
Neither of *n* and the gutturals surd and sonant as those combine with *n* to ɴ. Also *n* becomes *m* before a labial with one
exception of *Banba*, a name for Ireland, which is always
scanned trisyllabic in the poetry. Hence *n* makes svar. only
before guttural and labial spirants. Instances of the following
occur:—

$$l, r, \begin{cases} p & b \\ & bh \\ ch & \\ c & g, \end{cases} \qquad n, \begin{cases} & b \text{ in } Banba. \\ & bh \\ ch & \end{cases}$$

Examples: — *l, colpa* ᴋᴏʟəᴘə, *Alba* âʟəʙə, *balbh* ʙâʟᴜv,
dealbh daʟᴜv, *folcadh* ꜰᴏʟəᴋə, D. R. 94, *Salchóit* now 'Solla-
head' in Tipperary.

r, cuirpe? ᴋɪʀəᴘə, *fearb* ꜰaʀᴜʙ, *tarbh* ᴛaʀᴜv, *searbh* ѕaʀᴜv,
dorcha ᴅᴏʀəcə *(-rc-* wanting[1]) *fearg* ꜰaʀᴜɢ.

n, Banba ʙâɴᴜʙə, *banbh* ʙâɴᴜv, *Donnchadh* ᴅᴜɴᴜcə
seanchus ѕaɴᴜcis.

2. The svar. developed by *rr, n,* and in a few instances

---

[1] Except *orriric .|. oirdearc,* Cat. 12, be regarded as an example.

by *m* is sometimes used independently of the inducing context.
There is no lengthening before *rr,* or slurred diphthongization
before *nn, mm* in word pausa when the heavy ending is
resolved. *fearra dhuit* faʀə yot from făʀ, fắʀ, *ar a barra*
bắʀə from bắʀ, *gearra* gaʀə from gāʀ, *cama-chlis* kằmə from
kauм. *an-*=very *(ró-* always=too) is aнə- before all elements,
an-=un— usually combines with its word, *antoil* ꞌauнтḷ.

3.   Example of *gd* in *dearg daol (?)* daʀugə deʟ, of *ngbh*
in *diongbhail (?), diongabháil* O'R. *dingivalta* Cat. *-the* in the
ppp. of verbs is joined to a consonant auslaut root usually by
a svar. vowel, *ioctha* īkəhə, but *cortha* koʀhə, *curtha* kuʀhə.

Of the Consonants.

§ 44.   The voice-production in Irish is legato not staccato.
One might regard a continuous even voice or air-stream sub-
jected to the manipulations of an independent set of modifying
or articulating processes.   The air-stream may be considered
as a long irrational vowel interrupted and modified by certain
contacts, half-contacts and approaches.   Hence the perfect
agreement between vowel and consonant in timbre relation,
for the vowel is produced by the consonant.   Also the partial
agreement in intonation, e. g. ō with a nasal to ū, § 18,3, ə
to i before *s,* to u before a guttural, etc.   This intimate bond
connecting vowel and consonant and the persistent or continuous
character of the voice stream will explain the Keltic peculiarities
of voicing intervocalic tenues, or opening to spirants of unsup-
ported inter-vocalic consonants, of the tendency to anticipation in
sound formation which so helped the backward run of vowel
and consonant palatal umlaut in words, of the prevalence of
glides, of the phenomenon of auslaut hardening, the result of
a conscious effort to check the vocalic stream in pausa.   Here
glides are not reported except where (as in connection with
labials or gutturals) they have a specially prominent value.

Labials.

§ 45. The silent labials p, f, voiced b, v, w, and labio-nasal m, both broad and slender represent bilabial sounds. I regard labio-dental sounds as non-existent. Finck who does not appear to distinguish timbre in labials says that v and f are produced „zwischen der unterlippe und den oberzähnen,“ *Wörterbuch der westirischen Mundart,* p. VI. The upper teeth, a rigid element, are not used and so there is freedom for the production of broad and slender timbre. In regard to distance from teeth, rounding, or tension, the lips are by anticipation in position for the following vowel before the contact or approach for consonant production is made, and so broad and slender timbre can be at once distinguished. The former is produced with rounded, soft, protruded lips (as when one with lips held in position for ŭ makes the consonant contact for ᴘ), the latter with lips drawn tight, close to the teeth and inturned (as in the *ü* position). Hence the very wide difference between the m sounds in *ainm* and *anam,* the v sounds in *a mhic* and *fuinneamh.*

p

§ 46,1. *p* = ᴘ with a broad vowel. *Piarus de Poer* ᴘᴇʀ, 'Pierse Power', *parrthas* ᴘᴀʀəhis, *parr-* from *parad-* and *-th* with svar. vowel developed after *rr*. (For assimilation cf. *carrghios* from *quadrages-,* *orriric* .l. *oirdearc,* Cat. 12). *pobul* ᴘᴏʙʟ'.

2. *p* slender = ᴘ. *pic* pik, *preabadh* pr'ᴀʙə, *peacadh* ᴘᴀᴋə, *piuc* ᴘʏᴜᴋ, a morsel, also the sound made by chickens that have eaten dry meal. *seilp* Eng. 'shelf' an instance of auslaut hardening. D. R. 90.

The *s* of *sp-* resists palatalization, *p* is according to the vowel, *speal* sᴘᴀʟ gen. spelə, *spiorad* sᴘʀ'ɪᴅ.

p does not occur in auslaut except in such loan-words as *poimp* Eng. 'pomp'. In *stumpa* sᴛᴀᴜᴍᴘə Eng. 'stump' and several others a vowel is added to avoid the unusual auslaut. *p* is sometimes assimilated after *m*, *campaidhe* KAUMhī, camps, S. R.

4. A secondary *p* seldom written is produced by contact of *b* with the *h* relic of *th*, *ch*, or *f* (in fut. and cond. active) so *b*, *d*, *g*, and *v* are provected to tenues in the fut. and cond. through regressive action of *h* from *f*. (This *f* though invariably written is not pronounced in the active voice except it be the outcome of such groups as *bhf*, *mhf*, *bhth*.) The same rule has been in operation during the whole course of historical Irish (e. g. *intathair* from *ind-sh*) and is still active.

f

§ 47,1. *f* broad = F. *faire* Fȧr'ə, *fána* Fȧ̊nə, *faithche* Fȧhə, *fuair* Fūr' and hwūr'.

2. *f* slender = f. *fear* fȧʀ, *fir*, fir', *feall* fyauʟ, *fionn* fyauɴ.

3. *f* like all tenues is voiced by *n*. *go bhillfidh* ɢu vilhig, or with a pronoun ɢu vilhə.

4. *f*(= Fw, F, and f) becomes null on aspiration. *go lá fheil* ɢu ʟȧ̊l, *oidhche fheil* īhl, *mo fhir* mir', *mo fuiy* Eng. 'whip' mip, *fuil* Fuil, aspirated il, so F even with w glide disappears. This glide in often the first element of a diphthong consonantized. Accented or sound-bearing elements however remain after a labial. *buail* ʙūil, *fuar* Fūəʀ.

5. In O. I. there is observable a change in writing between *b* and *f* which is often further complicated by a pronunciation *h* in modern Irish. This change to *h* is in peculiar contrast with the ordinary behaviour of *f* or *b* under aspiration. Vid. Pedersen, *Aspirationen i Irsk*, p. 19. Thus: *beos*, *fós*, *fa* for *ba* in Keatynge, *féin* = *bé-sin*, Z². 366, Dēsi *féin*, *héin*, with *féineach*, *héineach*. *fuair* Fūəʀ but *fé fuair sé é* fē hwūər' sē ē, before he got it, S. R. *ar fuaid* er hwūiᴅ, *mearbhal* maʀəhʟ', *tinneas gan branda cheann is mearfoll* 'head-ache without brandy' *Uilliam Dall* (Tipperary)

Ren. 69. p. 40. *mearadhbhaill,* Fitzgerald Poems, Ren. *tairbhe,*
O. I. *torbe,* ᴛᴀʀəfə. The word ɢᴀʀəbɴ' is deduced from
garbhfhonn 'rough land' by Mr Carmody of Comeragh Mills.
anfadh, W. *anfud, fri ainbthe; M. ní Dhonugán* in O'Neil
Ms. *ainighthe,* with the usual svar. form of *an-,* § 43,2, a
storm, a great blowing, *meanfach* miáɴūc, $f = bh$ vocalized
after a liquid § 10,7, yawning. *tafann* ᴛᴀʜɴ', Com. song,
tafaint Dēsi, the barking of dogs and so hunting away with
dogs. *anbruith* aɴɾʜə broth. The old group *sv* gave in turn
s, b and *f. dosennat, dosephain, tafnetar* all from the same
root. Vid. Thurneysen, Index Z². So the forms *siur fiur,*
sister, *mórseser* and *mórfeser* seven men, Dēsi ᴍūʀ-hesɾ also
from *sv. taisbénadh* in Keatynge is pronounced sᴘånt, 2 per.
imperative sån in the Dēsi; *as-fenim* testificor, *taiss-fenim,*
Ich zeige, weise, führe vor, W. in voc. Also *taisbentar.* It is
interesting to note that the Eng. word 'Faith' used as an
exclamation in speaking English is pronounced 'hēᴛ in the
Dēsi. *acfuinn* is åᴋɪŋ with perhaps a lost *h* after a tenuis.

6. *f* in fut. and cond. as already mentioned becomes *h*
in the active voice. This appears as an aspirate after vowels
and liquids and provects mediae to tenues. After a tenuis it
appears to become lost, for an aspirate after a tenuis is very
weak or silent. An *f* is often restored by conjunction of *h*
with some relic of the auslaut consonants of roots. Examples: —
chidhfeadh cihᴜc where *dh* is merely ornamental, *leanfadh*
leanhach, Cass. 259, *leaghfadh* may melt *liefeach,* ib. l'yåhᴜc
Dēsi, *gh* + *h* represented by *f* in Cass. *daorfar e,* passive, ib.
135. *thabharfas* (instead of *dobéra) hourhis,* ib. 143, *graifigh*
tu ib. = *gráidhfidh* for *grádhóchaidh tu, mairhigh tu,* ib.
traochfar ᴛʀᴇcғʀ', S. R. ner ə hēdfʀ' = *nuair a shéidfear,*
Crowley. So ᴋailhɪŋ I would lose, *mairfeadh* ᴍårhᴜc he
would live, unvoiced r. ᴋɪrhə sē he shall put, *ní fhanfadh*
nī vaɴhu'c from *ní dh'fhanfadh, muna dtuillfinn* ᴍårə ᴅilhɪŋ,
(lh makes position for the production of î) *fanfaidh* ғåɴhə,

bhfuigbhthá 2 per. cond. vîFắ, ᴍắʀə vîꜰʀ′ M. song, *chifer*, Denn 83. *leacfar* ib. from *leagaint* to cause · to fall, *do dhiúgfadh cáirt* ə yūᴋᴜᴄ ᴋắrt who would drink a quart, M. song. ɢᴜ rēkə ᴛᴜ̄ from *réigh* ib., *ní thréigfinn* nī hrēkiꞃ, ib. *ní bhrisfinn* nī vrisiꞃ, yaɴꜰắ 2 per cond. but yēɴhiꞃ 1st per. I would do. This -ꜰắ in 2 per. cond. comes from O. I. *-ftha.* yīʟhiꞃ, cīrhiꞃ hīnhiꞃ yīsiꞃ S. R. *dhiolfainn, chiorfainn, shínfinn dh'iosainn,* I would eat. nī cimāᴛᴜᴄ Com. song. from *coimead, raobfainn* ʀᴇꜰiꞃ, Bob, *scriobhfainn* skr′īꜰiꞃ, *teibfeas* tepis, *osgail an dorus nó leagfam istig aguib é,* laᴋᴍ′, my Father, *dá ndeunfaidhe,* ắ nīaɴꜰwÍ, *go me fa,* Dunne = *go mbeidhthá, shuihegh* = *shuidhfeadh* ib. *dá siubhailfinn* ắ sūlhiꞃ, ắ sūlōiꞃ S. R. ꜰắkə ᴛᴜ̄ thou wilt leave, from √*fág-, cá bfuighthi i* ᴋắ vîꜰuÍ Í, where would she have been got (otherwise).

7. *f* is often prothetic. *taob na faille, fiarfug* fīʀhig 2. imperative, *fathach* ꜰắhÍᴜᴄ. = *athach.*

8. *f* outside the case of fut. cond. of verbs arises from *v + h. liomhtha* līꜰə, *neamhdha* nắꜰə 'heavenly' probably on the analogy of *naomhtha* ɴᴇꜰə holy, as elsewhere *-dha* of adjs. becomes -ɢə, *sgafaire* from *sgathmhaire* = *h + v.*

b

§ 38,1. *b* broad = ʙ, slender = b. *ball* ʙᴀᴜʟ, *bóthar* ʙōhʀ′, *blosgadh* ʙʟᴏsɢə, a flush, O. Eng. *blyscan* 'blush', *bárr* ʙắʀ and ʙắrə, *baoghal* ʙwᴇʟ, *buidhe* ʙwÍ.

2. *b* slender, *bith* bí, *ar bith agat* er bihᴇɢᴜᴛ, *breagh* br′ắ and br′ā, *beo* byō, *binn* baiꞃ.

3. *b* in auslaut broad and slender *badb* ʙîʙ, *fadb* ꜰîʙ, *ladb* ʟîʙ, *binib* binib, *breib* br′îb Eng. 'bribe'.

4. *b* assimilated, *diombáidh* diᴍắ, *domblas* ᴅᴜᴍəʟis, gall, D. R. 62.

5. *b* affected in anlaut = w, v, mostly w before a broad consonant, v before a slender. *a bhuachaill* ə wᴜ̄ecḷ, *a bhaile* ə wắlə, *a bháis* ə wắs, *a mhac* ə wắᴋ, his son. It is often

difficult to determine whether w or v is heard. *ar a bharra* er′ ə vârə, or wârə. *a bfuil* when *a* unaccented is silent = ʙul. *bhur* = O. I. *barn* wŪʀ or Ūʀ.

bh in inlaut between broad vowels = w, and contracts with its vowels to ou, § 5,1, unaccented to Ʊ. *abhainn* ouiɲ, *adhbar* ouʀ and āvʀ′ S. R. an earlier stage of development preserved in a story, (in *fádhbhar* ғāvʀ′ *dh* is a length making device) *cobhair* ᴋouʀ′, *leabhar* louʀ from *liber* where *e* was regarded as a broad vowel unless the derivation is from *librum* liʙʀ′. *labhairt* louʀ′t, *tabhairt* τouʀ′t, *seabhac* souᴋ, *abhac* ouᴋ, a dwarf.

6. *bh* in inlaut becomes a w glide after Ʊ sound vowels or diphthongs, a y glide after ī sound. *siubhal* sŪwʟ, *dubha* ᴅŪwə pl. fem. of ᴅuv, *subhachas* sŪwəcis. After long vowels and diphthongs it becomes absorbed along with its indices, *aoibhneas* īnis, *riabhach* rīuc, *diabal* dīuʟ. In *cuibhe, cuibhreach, cuibhreann,* there was first umlaut to ɪ and later contraction with y of v to ɪ, ᴋɪ, ᴋɪr′uc, ᴋɪr′ɴ′.

7. *bh* under conditions set forth in §§ 6,1. 8,1, contracts its y to form î, ᴇi.

8. *bh* is sometimes sounded as v between vowels in songs. *tabhairne* τâvr̥nə, Eng. ʻtavernʼ, also *táirne* which should be τouʀnə, *cobhair* ᴋovr̥′, T. G.

9. *bh* = v unaccented after *l, n, r*, contracts when broad with a svar. from the liquid and the following vowel to Ʊ, slender becomes y and contracts to ī. *arbhar* aʀŪʀ, *iorbhas* iʀŪs, *searbhas* saʀŪs, *marbhadh* ᴍâʀŪ, *carbhas* ᴋaʀŪs, as if phonetic for ʻcarouseʼ, *miorbhuil* miʀŪil, *ciorrbhadh* ᴋīʀŪ *dearbhadh* daʀŪ, *Cearbhal* ᴋaʀŪʟ. *banbh* pl. *bainbhe* ʙânî *leanbh* laɴuv but *leanbhin* laɴīn.

10. *bh* in auslaut = v, v, or silent. *Sadbh* sív, *banbh* ʙâɴuv *marbh* ᴍâʀuv, *craobh* ᴋʀᴇv, *sliabh* slív, *cliabh* klív. v in *taoibh, leinbh, fearaibh,* etc.

11. *bh* in *-ibh* of dat. pl. is mostly silent. However *fearaibh* is the Dēsi ,pl. in all cases. *air beoibh — air mairibh* Cass, ᴎå glasiv .|. *ná i gcleasaibh* Crowley, *Uibh Eirc* Iᴠᴇrk a place name, *d'annamnuiv*, cat. 11. *groihiv* ib. 14.|.*gcroidhthibh, a dtri parsaniv*, ib. 19, *ar veoiv* ib. 20, *ansna vireniv* .|. *bfiréinibh* ib. 21, *dar ngniorhiv*, ib. 23 .|. *gniomharthaibh, o varuyiv* .|. *mharuighthibh*, ib.

12. *bh* is often absorbed after a long vowel in auslaut; *maidhm sléibhe* ᴍîm slē, *neascóid cléibhe* a boil on the liver naskō klē, my Father, *Currach bhealaigh an tsléibhe* a place name in the Dēsi, ᴋʀâc valɴ' tlē or klē. So *Sliabh* in place names *Sliab Ruadh* slī ʀŪə, so *Sliab gcruinn* slī ɢʀaiɳ but slīv ɴə ᴍåɴ *Sliab na mban*.

13. *bh* is assimilated before a pronoun, *á raibh tu* ə ʀᴇ ᴛŪ, but *ní raibh* ʀᴇᴠ, lost in *tuarasgbhail* ᴛuʀɪsɢål, *faghbháil* ꜰåil. Assimilated in *is taolay ataoim=taobh lag*, Ren. 69 p. 8.

14. *bh*=g in *do sgriobh* sgrig, ppp. sgritə with usual shortening before *-te*.

15. The w from *gh* is hardened to v in auslaut of *ogh* an egg and *tiugh* thick. Also by the same hardening *andiu* developed ɳiuv, the voice-stream was continued until the lips had passed from rounding to spirant approach.

Dentals.

t.

§ 49,1. *t* broad and slender=ᴛ and t. As in the case of ᴅ, ᴛ is produced by pressing the tongue against the upper teeth; for t and d the tip of the tongue touches the hard gum about a quarter of an inch above the roots of the upper teeth and contact is so gradually broken that a slight spirans may be heard after the consonant. This spirans took the place of the consonant in western Irish, hence t, d is there spoken as *ch, j* in English.

2. *t* in anlaut; *tabhairt, tormus* ᴛoʀᴍ'is, stubborn pettishness.

3. *t* in inlaut was preserved only in consonant groups
otherwise it became *th=h*. (An affection is sometimes found
after *l*, *althóir*, *reulthan; toilhineach*, Cat. 47.) *slat, geit*, a
start, *siotadh* a .whinnying, its derivative *siotaireacht* crying
without tears (of a child craving something), *fíotán* a stiff glass
of whiskey are examples of old double *t*. Single *t* is found
in inlaut in loan words received after the working of the
vocalic infection *peata* a pet, *pótaire, poitín, pratainn* a 'patent'
with infixed *r*, *cúitiughadh* 'requite' Pedersen. Perhaps rather
from the form 'quit'.

4. The group *-cht* is unaffected and always broad. The
palatalized *boict* of O. I. and found also in Keatynge was
merely a symmetrical writing.

5. *t* after *s* in inlaut=*d*, in auslaut again hardened to *t*.
s causes exactly the same change in the case of *g, c*.

6. *t* as also *d* is often preserved in anlaut after a
foregoing *s, cionnus tá tú?* but *thá sé, is do bhí acu*. This is
part of the phonetic rule that *d, t, s, l, n*, and *d, t, s*, meet
in compound formation without disturbance.

t is written *d* in *dara* ᴛᴀʀ�.

7. The prep. *ind-* unaccented form *inn*, being accented
in noun composition took the form *int-* in combination with
the *h* relic of *s*. Thus *intshamhail* from *ind+h*. From this
went forth a sort of sandhi *t* between *n* and *s* in composition,
and later between *l, s* and *s*. Examples: — *intseachanta*
'vitandus'. (Here there is some doubt of the preposition though
it seems as if the new participium necessitatis was compounded
with the *in-* particle of *inairm* 'fit to bear arms', etc. not with
ind-). *caol t-sruth, le crainn tsleadh fhaobhar*, Denn, 80, *cois
t-Siúire* by the Suir, *min t-sruth, Cion t-sail* P. P. 312, *caomh
t-suairc*, O'Neil, *fionna-bhean t-séimh*, ib. *maidin t-samraidh,
S. na Sróna, milis t-séin*, T. G. 6, *an chlaoin t-saoghail*, ib.
31, *min t-sruithe*, ib. 44, *caol t-sruith*, ib. 95, *glas t-snuim*, ib.
35, *gleagal t-suairc*, Denn, Ms. Waterford College, p. 34, *buan*

t-seasamhach ib. *gein t-sleachta, Uilliam Inglis* in Ren. 76, p. 18, *an tsaoghail t-sogluaiste,* ib.

8. *t* appears after *n* final of verbal nouns, after slender *s* of 1st and 3rd per. pl. of imperative and conditional, also sporadic after *-is.* Its presence in verbal nouns may be traced to analogical or phonetic causes. It may have been added on the pattern of *tabhairt, labhairt,* or the alveolar *n* by breaking contact gradually during continuance of the voice stream developed the final consonant. Even so alveolar *s* may have produced it by making contact in pausa.

Examples: — *feuchaint* fīàcṇt, *gabhailt* ɢōlt 'going', *crádh-chaint, leigint* liɢṇt with *i* umlaut, *foirithint* helping, Dunne, *foireachtaoin,* S. R. *taosgaint,* Dunne, *cosaint, insint, seachaint* sàcnt, *feodhchaint* fyōcṇt, *grádhchant*Denn 68, *tarraingt, fanamh-aint* fᴀɴŪnt, *fuirist* from *fuiris d-uil* and *d* to *t* after *s* in auslaut § 49,5, *aithint, dóthaint* ᴅōcṇt and ᴅᴇcṇt, *go dtioc-faidist* ɢᴜ dᴜкɔdīst, *suibhcant, M. ní Dhonagán,* in O'Neil.

9. *t* from *d* in *t'athair* ᴛahṛ' or hahṛ', *led' t fhior-fhuil.* This is a revival of old *t* under the accent. *th'ainm* hanṃ, *ad' uachtarán* ᴛŪcᴛʀ'ᴀ̃ɴ, *t'irein*=*d'fhireun,* Cat. 3. Cf. O. I. *do-ind,* accented *tind-.* *t* provected from *d+h* in *ceudta* hundreds kēᴛə, sing. kīᴀ́ᴅ.

10. *tl*=*cl* . *t-slighe* klī, so *dl*=*gl, dligheadh* glī.

11. *t* in *acht* is lost, àc, or nàc generalized from some *n* auslaut.

12. *t* assimilates *c* in *Portcládhach* ᴘᴜʀᴛʟᴀ̇̃c, Portlaw.

13. *th* = h. *thoil* hel, hil, hoʟ, *athair* ahṛ', *ceatha* kahə. In auslaut *th* makes a short vowel explosive and appears as weak h before a following vowel. *bith, bih* Cat. 27, *caith uait ē* кà hwŪᴛē. *is maith é sín,* is ᴍà hē sin. Apparently silent after a long vowel sound, *treith* tr'ē, *díth* dī.

14. *th* broad in auslaut of the nom. case of nouns of the vocalic declension which make an increase in the gen. is hardened to c.

Examples: *rioth* rʋc, *i* umlauted to ʋ by the dark timbre of a guttural, gen. rahə, *gaoth* ɢɪhə, *luach*, Denn 78, = *luath*, *breach* = *breath* ib. *cleath ailpín* a fighting wattle klac, gen. klehə, *dath* ᴅàc, gen. ᴅàhə, *cioth* k(y)ʋc gen. kahə, *sruth* sʀʋc, *sruthán* sʀʋcàɴ gen. *sruithe*, *crioth* kr'ʋc gen. *creatha* kr'ahə, *cruth* ᴋʀʋc, gen. *crotha*, *guth* ɢʋc, gen. ɢohə. So *sgáth*, C. M. O. 34, *gruth*, ib. 102, *rath* ʀàc O. I. *rad* gratia, *ag brath*, P. P. 24, *liath* gen. *léithe*, *bráth*, *sgiath* scutum *scēt*, now a hemispherical basket used in picking or straining potatoes, *leath* lac, half, but *tair i leith* ᴛar ə le, *bláth*, *bláth na sgeach* ʙʟàc ɴə sgàc, *rathmas?* in good case, wealthy, from *rath?* ʀàcᴍ'is, *tráth* ᴛʀàc, *eadarthráth* àᴅr'ʋc dinner-time, but *virah=mhiorath*, Cat. 9, perhaps *h* was phonetic for c here.

15. In other auslauts is has the weak sound of *h* mentioned above *tnúth* ᴛɴꝋ, *tláth* ᴛʟå. (Here ᴛʟ is distinguished from ᴋʟ by the ʟ which in the ᴋʟ combination is sounded with more resonance room in the guttural chamber), *cath*, *fuath*, *dlúth*, etc.

16. *th* in inlaut between vowels = h. *athair* ahɹ', *tláthas* ᴛʟåhis, in Thom. contracted *tuille gan tlás* C. M. O., *áthas* åhis, *bathas* ʙàhis, *lahar*, Cat. 9. So ppp. *cumtha* ᴋʋᴍəhə *foluighthe* ꜰoʟɪhə.

17. Adverbs of place which show an anlaut change between *th* or *sh* and *s* use the *th* or *sh* forms to denote rest in a place, *s* motion to. Motion from is denoted by the anlaut *an-*. *thoir* hɪr (or hir as h does not affect the vowel timbre) *thiar* hꟾR, *thuaidh* hꝊig, (one of the many examples of auslaut g fondness), *theas* has, in the East etc. *chuaidh se siar* hwꝊ sē sꟾR he went westwards. *fó dheas* is ō yas. *sall* is not found and *anann*, *anonn* ɴ'aʋɴ, is the reverse of *anall*, *anann* to yonder, thither, *anall* from yonder, hither.

18. An h appears to be developed occasionally after *rr*, probably the stroke of the vocal chords being again set in vibration after unvoiced *r*.

d.

§ 50,1. *d* broad=ᴅ, slender, d.

ᴛ : t=ᴅ : d. The production is the same as for ᴛ, t in all respects except for the weaker explosion and the voicing.

d occurs in anlaut *dúmas (?) athá me* ᴅŪ̄ᴍis, I am only joking, *dúmas go raibh sé tinn=dubh-amus? ag dul* ᴅᴏʟ, in Connaught assimilated to *g*, ɢᴏʟ. In the Dōsi *d* does not assimilate with *g*. One instance *ba bhreug dam*, ʙᴏ vʀēɢᴜᴍ M. song, perhaps of Western origin. It is remarkable that strange dialectic forms are preserved in songs and stories, each person repeating the words as he learned them and not forcing them into conformity with his every-day speech. *dubhairt* ᴅŪ̄rt a rare instance of ū after a broad consonant, *dada* and *dadamh* ᴅᴀᴅə, *tada*, OR. *gan ttada do shuim mar leanabh* R. I. A. 52 p. 259.

2. In inlaut the result of worn down groups *adeirim* from *at-bh-, admháil, adubhairt*. Preserved after *r*, *árd, bárd bórd*. In inlaut after *s*, *brosdugadh*. The group *st-* becomes *sd* in inlaut, in auslaut it is hardened to *-st*. In auslaut there is sometimes indecision after *r*, *cuaird* a visit, an old loanword. The same word borrowed again later is *cuairt* 'a court'. *magcuairt=imman cuairt*, with *gc* writing of the tenuis voiced by a nasal, has *-rd* or *-rt* in auslaut. So *d* represents *t* voiced by a nasal in *cadtrom* ᴛ́ᴀᴅʀ'ᴍ from ᴛʀᴏᴜᴍ, *eadtarbhach, idir* and *'dir, féidir meud. nead* from **nizd-*, by unvoiced umlaut provected to t over *r* by *h* in *eatortha, edar-h* to *etor*. In *seanad senāt-, sgrudadh scrut-, spiorad spirit-*, appears the usual change from tenuis to media in Latin words that were borrowed after the working of the aspiration law.

3. *agad* in sometimes written for *agat*, so *d* for *t* in auslaut after *s*, e. g. *arisd*, but after unvoiced *s* it is impossible to sound *d* except a voiced element follows.

4. *id'* 'in thy' is always *in do* before a consonant in the

Dēsi. So also the other prepositions with *do, aige do mháthair*
for *agad mh-*.

5. *d* assimilates with *l, n*, and occasionally *r. ceudna*
kīáɴɔ, *órdlach, codladh* ᴋᴏʟɘ. *a chodlfadh* a new conditional
ᴄᴏᴜʟʜᴜᴄ, Bob, where *h* made position because being a non-
vocalic addition the group *ll* was not divided § 2,2 VI. *carr-*
gios, quadrages- has already been mentioned, with k in *neascóid*,
cléibhe.

6. *d* is made unvoiced or provected to a tenuis by *f=h*
in fut. cond. *go ngoidfidh me* ɢᴜ ɴɪtɘ mē, S. R.

7. *dh* broad in anlaut=ɢ=*gh* broad in the same situation,
dhá ɣȧ.

8. *dh* slender in anlaut=y which is also=*gh* slender.
a dheor ɘ yōr´.

9. *dh* slender in auslaut is often hardened to g. *dhiaidh*
yīg, *suidh* sɪg, verb. noun *suidhe* sɪ, *luigh* ʟɪg, verb. noun
ʟɪ, *garsún cruaidh ón buaile aitinn* ᴋʀᴜig, *fáidh* ꜰȧig, *aghaidh*
ayg, Cat. 3, Dēsi îɘ, *réidh* rēg, *thuaidh* hwᴜig, in fut. and
preterite endings without a pron. *caithfidh* ᴋȧhig, but *caithfidh*
tú ᴋȧhɘ ᴛᴜ, *bhfhuighbidh* vîg, Walsh, *beidh* beg, but *beidh sé*
be sē, but *cruaidh-cheart* ᴋʀᴜᴄaʀᴛ, y lost after a long vowel
sound.

10. *-idh* as ending of 2 pl. imperative = ɘgī, Des. -ig.
buailidh ʙᴜlɘgī or ʙwelɘgī. *so* as in *so dhuit* sᴏ yet, takes
this ending sometimes in the pl. sᴏgī yīvsɘ = *so dhibh-se.*

11. The prep. *de, dhe,* generally after a vowel ye, yɘ,
has its y occasionally hardened to g, either a wrong restoration
for *d* or an instance of *g* fondness in supposed auslaut. *ráithe*
go leith de laethantaibh agus ráithe go leith d'oidhcheantaibh
ach trí ráithe glan ó san amach. ʀȧhɘ ɢᴜ legɘ ʟᴇʜɴ´ᴛɘ ȧɢis
ʀȧhɘ ɢᴜ leg īhɴ´ᴛɘ, Ph. H. *cúpla púnt ye sin,* S. R.

12. *dh* in the ending of the preterite passive = ɢ. *crosag*
T. G. 6. This rule is universal except occasionally in songs
imported from other dialect territories. *leagadh in cladh uirthe*

laɢuv ə ᴋʟĪ erhə my Father, reporting a Kerry conversation. This is the change from guttural to labial in the treatment of a consonant relic observable in such examples as *sawan*, *sehen*, etc.

13. *dh* in the adj. ending *-dha* = ɢ. *mórda* ᴍŪʀɢɔ, *órdha* ōʀɢə, *Chlíodná mhaorga*, S. R. So *riogach*, *cróga*, *diaga*, *yiaga*, Cat. 7, *diaganta*, *anvasarga* Cat. 7, = *aɪnmheasardha*, after *ll*, *gallda* ɢauʟᴅə, after *m* in *iomdha* ᴍŪ. In pl. of a noun *suadha*, *na suaga*, *U. Inglis*, Ms. in Waterford College.

14. *dh* is often inserted to represent a glide between two vowels. *clúdhamhail*, *miodhadh* mĪyå̃. Also for the same purpose when there is a change of timbre *beidheadh* bec, *bidheann* bĪyɴ', *do chidheann* T. G. 8.

15. *dh* which gave y and produced a slurred diphthong with certain vowels or influenced by final double liquids is often written to represent the y developed by double liquids when they produce the same sound alone. *greidhean* ɢʀ'ᴇɪɴ, so analogically *greidhm* for *greimm* gr'ᴇim. *coidhill* = *coill* ᴋîl, Ren. ʀᴄ. p. 65. *sgridhig* sgrĪg, ib.

16. *dh* from the analogy of so many silent *dh* auslauts is sometimes written ornamentally, *daoineadh*.

17. *dh* (and *d*) unaccented often falls out. *fear a bfearr*. So the prep. *do* before verbal nouns became ə; *do mharbhadh* ə vāʀŪ. *dia n-* becomes *dá* with eclipsis of following consonant and *d* falls out *dá ngrádhann tu ghlacadh* å̃ ɴʀå̃ɴ ᴛu yʟåᴄɔ, where also *do* is entirely lost before *glacadh*. So *oc a mharbhadh* = å̇ våʀŪ, *dá rádh* å̃ r'å̃, *tabhair dam*, Thom. ᴛōʀᴍ' (Dēsi ᴛouʀ) and shortened to ᴛoʀᴍ by vocalic break of pausa. Cf. *gablhtha* ɢoғɔ from ɢōl(?).

18. *dh* in *-adh* ending of verbal nouns is silent. *bualadh* ʙŪəʟɔ; this ə contracts with y, w, of *gh*, *bh*, to Ī, Ū after a liquid. § 10,7.

19. *dh* as ending of 3 per. sing. of imperative, imperfect and conditional=c. *biodh sé* bĪuc sē̄, *do gheobhadh* yōc.

20. *dh* is silent after a long vowel sound in syllable auslaut. *buadh*, ʙ∪ə, so *clódh, grádh, crádh, gruaidh*. In such instances as *bliadh-ain glaodh-uch* it resumes a glide value. So *adh-bhar* ăvʀ', ouʀ, *Adh-amh* ăv. In *riabhach, diabhal* there is a parallel loss of *bh*. ʀīuc, dīuʟ. Of course with short i we should have had ʀouc douʟ, § 5,1; hence the Western pronunciation of *diabhal* jouʟ goes back to di-, in the Dēsi Lat. *ia* became the Irish *long* diphthong īə, *ádh* ā, luck.

21. In *cladh* ᴋʟɪ, the writing from middle Irish *clad*, the pronunciation from *claide*.

22. *dh* in pl. ending -*adha*=ī. *curadha* ᴋuʀī.

23. *dh* (also *gh)* in accented syllable surrounded by short voiced elements gives î, ᴇi, §§ 6,1. 8,1. *adharc, tadhg, meidhg,* O. I. *tredenus* trîɴis.

24. *dh* and *gh* lengthen O. I. *e* through a to ā, ẳ. *meadhchaint* mācn̩t, Thom. ẳ : *sgadán,* D.R. 44, O. I. *med-ón* māɴ, but mōɴ : *brón : sóirt* Ren. 69, p. 6. *cned* knẳ : *bláth* T. G. 7, *sned* nits, snẳ, *spleadhchas* splẳcis, *bregda* br'ẳə, br'āə, *leaghadh* lyẳ melt. As *dh* lengthened e to ā here so must it have made a to ẳ thus O. I, *mad, mád,* modern ᴍẳ, *ádh* ẳ, etc.

25. *dh*=y contracts with i to ī. *croidhe* ᴋʀɪ, so -*idh* the termination denoting an agent ī. Also *gh, slige* slī.

26. *dh* with accented o and following voiced element=ou. *bodhar* ʙouʀ, *odhar* ouʀ, *modhamhail* ᴍoul § 5,1.

27. *dh* in auslaut before a slender vowel=y, *budh é* ʙu yē, *budh eadh é* ʙu ya ē, ʙuyin ē=*budh shin é, nár bhudh é an sganradh é* ɴẳʀə yēɴ sɢauʀ ē. Here *budh,* bad for old *ba*.

28. The prep. *do* (before a coloured vowel *d')* prefixed to verbal nouns with a slender vowel anlaut becomes y, the root is regarded as beginning with *do* i. e. *dh,* and another *do*=ə is prefixed. *chum é sin do dh'éiliomh* cuɴ ē sin ə yēluv. So in all verbal uses of *do* in the imperfect preterite and conditional. Also with *dá, dá dh'eiliomh* ẳ yēluv. *d* is however kept before *iarraidh, cuaidh sé d'iarraidh braonín beag bainne* c∪ə

sē dīr ʙʀᴇɴīɴ bᴇɢ ʙàɳə ʽhe went for a sup of milk', *bhi sé d'iarraidh a choimeud isteach* vī sē dīr ə cimāᴅ istàc. Here *do=ag.* The case of *dh* before a broad verbal anlaut is exactly parallel, *do dh'órdughadh* ə ʏōʀᴅū. *do=de* is sometimes treated in a like manner before other parts of speech *d'aon chloich* ʏᴇɴ ᴄʟᴏᴄ T. L. *ráithe d'aois* ʀåhə ʏīs, M. song, slender before a broad anlaut in *mála d'ór* ʏōʀ=*dhe ór* S. R.

29. The *t* of *tar* as oath formula = ᴅ. *tar Múrtin* ᴅᴀʀə våʀtɴ´, or null, *ar a bfulaing*, T. L. *ar ndoigh* ᴀʀ ɴū, § 18,3. So *tada* is *dada. dar fiadh* is ᴅāʀ fī, with a curious lengthening of *a.*

Gutturals.

c

51,1. *c* broad ᴋ, slender k. Affected broad becomes a spirans ᴄ, slender the spirans ᴄ.

Occurrences: — *coll* ᴋᴏᴜʟ, *corcor* from ʽpurpur' of the early British missionaries from the analogy that *c* Irish was = *p* Welsh. Inlaut and auslaut the result of a group or provection. *leac, pluc, sprioc* sᴘʀᴜᴋ, a spurt, a start to work, *deacair, craorac.* Slender, *cill* kîl, *lice* likə, dîʀik a mountain name near Mount Melleray in the Dēsi.

2. *ch* broad in anlaut and auslaut = ᴄ. *mo chreach is truimme mo chroidhe ná cloc* ᴍᴜ cr´ac is ᴛʀɪᴍə ᴍᴜ cʀɪ ɴå ᴋʟᴏᴄ, so in auslaut *geallach* ɢᴀʟåc, ɢʟ´åc, *marcach* ᴍᴜʀᴋåc gen. ᴍåʀkig. In *cht, beannacht* bɴ´àᴄᴛ, *ocht* ᴏᴄᴛ.

3. *ch* slender in anlaut ᴄ. *mo cheann* ᴄᴀᴜɴ, *a chleath* ə clac.

4. *ch* in auslaut parallel with *th* i. e. a foregoing short vowel is made explosive and an *h* is carried to a following vowel. *deich* de, *deich aguinn* dᴄhəcɪɳ, so *amuich* əᴍᴜ, after long vowel sound *fruoich* ғʀɪ, gen. of *fraoch* ғʀᴇᴄ *i gcrith*= *gcrích.* In inlaut after accent also = h. *fiche* fihᴄ, *da fhicid* ᴅᴀhɪᴅ, (slender d dental in auslaut), *duithche* ᴅūhə, *faithche* ғåhə, *choidhche* cīhə, *oidhche* ɪhə, *dicheall, ar nihil, dihil,*

Cat. 14, *inchinn* inəhiŋ, so also *caillichín* P. P. 162, *siorraichín giobalach, croiche,* S. R. *sgeiche, sgeichín, creiche* gen. of *creach* T. G. 11, *teiche,* after long vowel *féiche,* in anlaut *cheana* hanə. So *ch* broad in some words, *gach aon cheann* ɢu hᴇɴ cauɴ, *gach aoinneach* ɢu hainə *gaẊeingye* Atkinson, *Trí biorghaoithe,* Gloss. *imtheochad* imōhiᴅ, imōᴅ, *droch-* ᴅʀo-, *chucha* cŪhə, *dorchadas* ᴅoʀəhəᴅis, *cidhe* O'Neil = *ciocha* kī, *a ci* Com. song. *nach* only before predicates, *ná* before verbs. *adeirim ná fuil, fearaibh nach é* men besides him, *adeirim nach eadh.*

5. *ch, c,* broad changes a to à. *geallach, creach, cearc, dearc.*

6. *ch* appears in anlaut of *co, cu* with suffixed pronoun. *chugham* cŪᴍ, *chughat* cŪᴛ, *chuige* cıɢə *chuici* cıkə, *chughainn* cŪŋ, *chughaib* cŪv, *chucha* cŪhə. Here there is a mixing with *chum* cɴ´ from *dochum. chomh* appears also with aspirated anlaut from some such confusion. *chomh dona* cŪɴ ᴅuɴə, with assimilated nasal, *chommór* cŪᴍŪʀ. The lengthening is due to some position before a consonant anlaut.

7. *ch* (= *gh*) becomes y in *rachthá* ʀɪ̆ꞓ̊ *rachad* ʀɪ̂ᴅ. *ch* slender to y in *comairche, cloch na comairche* ᴋʟoꞓ ɴə ᴋuᴍɟɪ̄ a curious stone in Mothil. It comes from the writing *commairge,* vid. W. in voc. Cf. for the pronunciation *suirghe.*

8. *c* in *cionnus* = ᴋ, ᴋuɴis, 'cindas pro ce indas', Z² 357. But the pronunciation comes from *co-indus.*

9. *c* = k in *coimcad* kimāᴅ.

10. *c* is assimilated in *cúigcead* ᴋŪkɪ̄áᴅ, in *Portcladhach* ᴘuʀᴛʟꞓꞓ from the similarity of sound of the groups *tl* and *cl.*

11. *c* is lost before the accent in *cad é an* ᴅĕɴ, *cad é an rud é* ᴅꞓ̄ʀəᴅꞓ̄.

12. *ch* is sometimes added at the end of a word *maorgach* = *maordha, féineach* = *féin* M. song, *léireach* = *léir.*

13. *ch* becomes *h* and provects *b, iomchor* auᴍᴘʀ´.

14. *c* provected from *g* in fut. cond. act. is often kept with *f* in the passive. *leacfar* laᴋꜰʀˊ T. G. 37.

15. *c* hardened from *g* in auslaut; *gaedhlic*, Ms. R. I. A. 52 p. 310, end, so in Port Erin Isle of Man ɢɪlk vaɴɴˊ „Manx“, *comhrac*, *craorac* from *caor-dearg?* *lochta*, Eng. 'loft' was borrowed while *-ft* was still *-cht*. So 'trough' still keeps its old pronunciation in Ireland viz. ᴛʀâc.

g

§ 52,1. *g* broad and slender occurs in unaffected anlaut. In inlaut and auslaut it is the result of groups, the voicing of *c* by nasal affection, or auslaut *g* fondness. *gabhail, goile.* In *go* O. I. *co*, *gac* O. I. *cach* it is softened before the accent. The accented from of *cach* .|. *cách* is however preserved. The same phenomenon is seen in the case of *d* before the accent, *led' thoil,* etc. *beag* from *becc* bᴇɢ, *cogar* ᴋoɢʀˊ from *con-cor.* Slender *geal* gâl, *bige* bigɔ. *an-* = *un-* gives ē with nasal affection of a following consonant. Before *c* the writing is *eug* i. e. regular eclipsis. Thus from *cóir* regularly *eugcóir, eigceart* D. R. 64, *eugcruth* is sometimes divided *eug-cruth* as if it were a compound involving the use of *eug* death. In *eugmais* iâᴍis, the writing of the particle is one torn from a *c* anlaut. It should be *eummais.*

2. *g* broad and slender is found in auslaut both accented and unaccented as hardening from various consonant relics. It is so prevalent and arises from so many various sources that it must be ascribed to a g fondness in auslaut.

In preterite passive ɢ from *dh. baluigheadh* bâʟĭuɢ, *rugadh* ʀuɢuɢ, and so for all such cases. Another change of *dh* to ɢ is that in adjs. with a *-dha* termination. This however is not auslaut hardening. g appears much oftener. From *gh; tig* dat. tig, *luigh* imperative, preterite without a pronoun, ʟɪɢ, *marcaigh* ᴍâˊʀᴋɪɢ, *imthig* imig, from *dh. beidh* beg, *suidh* sɪɢ, *claoidh* ᴋʟɪɢ, from *bh. faig*, C. M. O. 21, *sgriobh* sgrig, ppp. sgritɔ. So in dat. of many verbal nouns, *ag léimirig sa preab-*

arnaig, for *léimnigh* and *preabadh, plabarnaig,* Bob. In *luigh, suidh,* the old short vowel is kept because the consonant did not become y and contract. The verbal nouns are long LĪ, sī. *claoidh* and *sgriobh* are shortened by analogy. *ságh-sa* sǎg-sə I stuck, T. G. 8, Lǎg he shot, from *lámhach, laaim,* W., *aghaidh, ayg.* Cat. 3, îə, Dēsi.

3. *gh* does not become g but is mostly absorbed after a long vowel. *truagh,* TRŪə, *dógh* burn, DŌ, C. M. O. 10, *dóthim,* O'R. *fligh* C. M. A. 7, *brigh* br'ī. In some of those cases the lengthening is due to contraction with y before the g hardening appeared. So *tuagh, laogh, sluagh, sogh, rígh* RĪ gen., *liúig* D. R. 91, but lyū, Dēsi, for the long vowel.

4. *gh* in inlaut is absorbed after a long vowel sound. *saoghal* SEL, *baoghal* BWEL, *soghach* SŌC, *laghach* Lǎc. If originally short those two last examples should become souc, Lîc, though they may have been derived from *sógh, lágh,* when those were regarded as SŌ, Lǎ. As long sign; *pughdar* powder, C. M. O. 12, *hughda* hood, ib. 12, *fághnach* P. P. 182.

5. *gh* broad or slender after *l, n, r,* § 48,9, contracts w or y with the svar. thrown out by the liquid and becomes Ū or ī. *feadghaile* faⅅlə, *murrghach* MURŪc; slender, *bairghin* Bǎrīn, *eirghe* Eir'ī *carrghios* Kǎrīs, *doilghios, dolish,* Cat. 35, *coingioll, cuniol,* Cat. 35.

6. *gh* lengthens e in *breaghdha,* br'ǎə, *deaghthach* dǎhuc, T. G. 47, but *sleagh* sla, So u in *chugham,* etc., cŪм' elsewhere cUGм'.

7. *g* seems preserved in *glégeal* glēgL' and glēyL', *rigne* is Rin.

8. Sometimes *g* contracts irregularly in imported songs. *tógadh,* Dēsi TŌGUG, *tóg* Dunne; *rugadh,* RUGUG Dēsi, RŪG in quoting a passage from C. M. O. Sometimes revived from the writing of *saoghal, baogal,* often SEGL', BEGL' in T. G.

9. *gh* broad=w and slender=y join with various vowels

in the production of slurred diphthongs. *aghaidh* ิə, *logha*
LOU, *lou* Cat. 76, *feighil* fɛil. In *aghaidh* apparently broad
gh=y but the O. I. form was *agid* and *a* must have been
umlauted before the contraction. §§ 4, sqq. Here *gh* is treated
exactly as *dh*. *togha* TOU, but ppp. *toghtha* TOFə, and cond.
pass. *toghfaidhe* TOFĪ. C. M. O. 11. In one place of C. M. O.
toghtha : foghluim and therefore TOUəhə as if *th* was joined by
a svar. vowel. In *toghtha* w was provected to f by h. For a
similar shortening cf. TORM´ from TOUR.

10. *gh* broad in anlaut=Y, slender=y. So *dh* broad
and slender. *mo ghrádh do ghalar*, MU YRẚ DU YẚLR´ 'I wish
I had your complaint' *mar gheall air* MẚR YAUL er.

11. *g* is provected to *c* in fut. and cond. *ní léigfinn* nĪ
likiɳ, *sloigfinn* C. M. O. 13 SLɪkiɳ. So with *th, tógtha*, TŌKəhə,
with slender *c*. Auslaut hardening in *comrac, craorac, aisiog*
and *aisioc*.

12. *gh*=y falls away before h in -*ighthe* ppp. endings of
-*ighim* verbs. Also *g* is wanting in *tionsnainn*, Ren. 76 P. 16
end, *dá thionsgailt* ẚ housɢɪlt M. song.

13. *g* after *s* is voiced in inlaut, unvoiced by auslaut
hardening in auslaut. *sgairt* Sɢẚrt, *sgeul* sɢĪàl, but *tásg* TẚSK,
rúisg RŪsk.

14. *g* in *ag*=g or ge, suffixed pronoun form for simple
prep.

15. *gh* in *tiugh*=v. This was a spirans thrown off from
the lip-rounding of *u*. So *ogh* UV. Vid. Pedersen *Aspirationen
i Irsk* p. 15.

16. *g* arises from *c* under nasal affection. *i gcorp* ə GORP.
The occasional change of the prep. *de* to *ge* has been mentioned.

The Liquids.

§ 53,1. N the dental sound. The tongue is in the same
position as for T, D; sometimes it is strongly nasal *naomh* NEV,
where the nasality of v affected it by anticipation, so L in *lámh*.

2. In unaffected broad anlaut it is a strong dental, in inlaut and auslaut after a sonant element it becomes alveolar. In *bainion* ʙunɴ́ 'female', for instance final ɴ is produced nearer to the teeth than n but yet without touching them. In unaffected slender anlaut *n* is alveolar.

3. *r* broad and slender appears to have the same sound in unaffected anlaut, affected it has the reduced sound of *r* in the ending -*aire* and in such combinations as *pre- bre- gre-* etc.

4. *nn* slender=ŋ or ny, *andiu* əŋuv, so *n* in *neamh*. *l* occasionally becomes very slender through combination with j, *leaghadh* lyā malt. A different sound is that in *baile, meala, bealaigh*, the *l* of *leaca* seems to resemble the ordinary anlaut Eng. *l*.

5. The reduced *r*=r′, though a distinct *r* sound, is not far removed from a strong y. In producing it the tongue is spread and hollowed spoon-shape the tip brought near the gum behind the upper teeth leaving a slight passage. The stream of air is directed towards the hollow of the tongue and plays against the upturned tip which is probably set in vibration. The teeth are held slightly apart. In the groups *pr, br,* the tongue is raised and there seems to be one impact against the gum at the beginning of the *r* sound. In Kilkenny the tongue is not raised and the teeth are almost closed, hence this r became ẕh, a soft voiced alveolar affricate.

n.

§ 54,1. *n* dental=ɴ, alveolar n, *nn* slender=*ng* slender =ŋ, *nn* broad=ɴ, *ng* broad=ɴ.

2. *n* broad and slender occurs in all positions in a word and is pronounced except where assimilated by *l* or *r*, or where in nasal affection it is a mere index that the following tenuis is voiced. Under vocalic and consonantal infection though there is change of timbre it never loses its *n* character. For a discussion of the delicate changes undergone in those conditions vid. Pedersen, *Aspirationen i Irsk.*

3. *n* in some words darkens long and short *o* to .ʊ and u, §§ 18,3. 13,5.

4. *n* in anlaut, *na* nə, *naoi* ne; *nó* nʊ, *nathair nimhe* now. sometimes divided *an athair nimhe*, natrix.

5. Inlaut *conadh, ceudna* kīānə, *im bliadna* ə mlīənə this year, *conach* kunàc. Auslaut *bán, bean, cion*, kun.

6. *n* slender; *neart* nart, *ní;* double, *buinneán* bwiŋȧn from *bun* + y? So *taithniomh* taŋūv. The difference between *nn* = ŋ and *ṅ* + y was so slight that the sounds are interchangeable. Auslaut *sin* sin, but sun after a broad vowel, an instance of umlaut working forward.

7. *n* = r in *muna* màrə.

8. *n* assimilated with liquids except *m. aonrudh* erud, *sganradh* sgaurə, *bainrioghan* baurīn, *branradh* braurə Hiberno - Eng. 'brand' a tripod to support a griddle, a gridiron, 'things 'd be like iron gates med' S. R. *anlan* aulṅ'; *aindeár* aŋr'; *pónaire* beans, pʊŋɽə and pʊɽə S. R. from *póinre* by contraction. *áilneacht* àluct, *an rós* ə rōs, *míonla mánla* milə màlə, *eanlaidh* īàlə.

9. *n* in *aon* assimilates with *ch* broad. *ar aon chor* ër e cr', *nín aon chomhairle* nīn e cʊɽlə, S. R. *ar aon chuma* er e. cumə.

10. *n* = *na* owing to the svar. developed in *sean-bhean* sanəvan. So in *an-* = very, also *an-* = un in *ainbhfios* anəvis. So *seana-sloc*, but *seanduine* sandinə.

11. *n* in the prep. *in* does not usually fall out nor suffer assimilation in the Dēsi. *suim an dia*, Denn, *dúil an psailm* ib. *in lúb coille*, M. song, not *i llúb, in seambirín* in a little room, ib. *an t'oige* = *id'óige, in drom a céile*, S. R. but *d'fhan an cholan im dhiaidh* yàn ə colṅ' m yīəg, ib.

12· *n* is inserted to break the hiatus between preps. and vocalic anlauts of pronouns. This at present purely sandhi *n* or ν ἐφελχυστιχόν derives origin from the nasal auslaut of *ré-n, le n-ól* to be drunk, to drink = *le n-a ól*, so *lè n-áireamh, le n-ithe* with pronouns alone *le n-a, Cáitín ó n-a máthair, do n-a chéile*, S. R. *dá cheann* to his head, in songs, otherwise *do n-a cheann*.

13. A *-na* plural appears in *lámhna, cnámhna,* ʟâɴᴇ, ᴋɴâɴɘ which is but the nasal *mh* under a change of form. So *comh dona* ᴄᴜɴ.

14. *n* appears as deriving suffix in some denominatives. *breugnadh, mo cheusnadh* T. G. 27, *criochnadh* ib. 61, *ceasnach* P. P. 306. Those have a causative signification. Cf. *bertaigim* and *bertnaigim, crithnaigim* from *crith* in W. and numerous other instances in middle and even in old Irish. This *-naigim* ending may have been torn from some case of the use of *-igim* with a stem having nasal auslaut.

15. *nach* represents *ach = acht* occasionally. It is doubtless taken from some familiar phrase where it followed a nasal auslaut, such as *ní bhídheann ach. ní bheirim sa chuaird liom nach a chircín bheag suarach* T. L.

16. *n* of the article is sometimes kept, sometimes falls out. There seems to be no rule. *táinic an buachaill, 'n ghaduidhe* gen. S. R. *an córda,* ib. *an dara h-uair tu,* ɴ′ ᴛᴀʀɘ hᴜʀ ᴛᴜ, ib. *as in áit,* ib. *dín mhuineal = de + i,* ib. *ar am bróg* er ɘ ᴍʀᴏ̄ɢ, ib. *'gimtheacht an bhóthair,* ib. *casadh an bhróg ghlan air,* ᴋâsᴜɢ ɴ′ᴠʀᴏ̄ɢ ʏʟâɴ er′, ɴ′ ᴀᴅ is ᴠɪ̄ sē while he was, ib. *dhen lochta* ʏᴇɴ ʟᴏᴄᴛɘ from the 'loft' ib. (but *dín mhuineal,* also *do'n fear* ᴅᴜɴ faʀ) *fear 'n tanyard,* ib, *tabhairt 'n éitheach* ib. *raibh an t-action dá dhénamh* Carrick-Shock song. Without *n, mise fear a tighe-se, a bheirt* S. R. *ar taob a bhóthair, insa choill,* ib. *fear a tanyard,* ib, *bhuail sé 'sa lorga,* ib. *a cás,* ib. *a córda,* ib. but also *in córda,* ib, *dé chúis?* why? *atá = an t-ádh.*

17. *an* comprising the prep. *in* and the article causes aspiration in such phrases as *chuaidh tu an chill, Anna, táinic an buachaill an bhaile,* S. R. *cuaidh sé isteach in scrubarnach,* he went into the shrubbery, ib. *'steach in ti*

stâc in ti, ib. *ann gach tir* shows *n* of the prep. kept before *gach*.

18. *n* slender + y is almost ŋ. *an iubhar* ŋ yŪʀ, or ŋŪʀ, *andiu* ŋyuv or ŋ'uv; in anlaut more clearly ny, n͞eamh nyâv, breaking of *e* before a broad labial, but *neamh-* from *neb- neph-* na- or nav-, naf- according to the following anlaut.

19. *nn* broad and slender constitute heavy groups in auslaut. So single *n* broad or slender in position. *nn* broad developes anusvâra in this situation, *nn* slender gives y. Those induced consonants are fixed in position but on resolution of a heavy auslaut group, or with loss of accent, the original vowel is retained. *ceann, gleann* kauɴ, glauɴ, but gen. glanə, *cinn* kaiŋ, but *binne* biŋə, *clann* but *clanna Gaedhel* ᴋʟᴜɴə ɢᴇʟ. Svar. *fionna-bhroig*, P. P. 94, *fionna gheal* ib. 226, *fionna bhean t-séimh*, O'Neil, Keatynge, *ranna suilt*, P. P. 150, *fonn* ꜰᴏᴜɴ.

20. ɴ, ŋ. *thánga* håɴə and håɴ-ɢə, S. R. *teanga* taɴə, *seang* sauɴ, but *seanga-chuirp* saɴə-ᴄɪrp, P. P. 212, *aingceis* aŋis, *aingeal* aŋʟ', T. G. 4. Often in auslaut *túirling* ᴛŪrliŋ, *aisling* asliŋ, *fairsing* ꜰårsiŋ. Here ŋ is plainly heard [as also in inlaut before a consonant, *dámbeidhinn-se* å meŋsə. An anlaut ɴ, ŋ, arises from nasal affection of *g*. *i ngeall* ə ŋauʟ, *i ngabhal na croiche* ə ɴᴏᴜʟ ɴə ᴋʀɪhə, S. R.

21. *gn* = ɢɴ', gŋ. *gnó* ɢɴ'Ū § 18,3, *gnaoi* ɢɴ'Ĭ, or rather ɢɴĬ, *aigneadh* agŋə.

l

§ 55,1. *l* dental-guttural = ʟ. The tongue is pressed softly against tho upper teeth or gum while the guttural chamber is made wide as in position for a deep Ū. The sound can be best attempted by starting it from such a vowel. Its guttural character is attested by the fact that it affects vowels as the other guttural consonants, by its colouring the irrational vowel to u, and by the fact that children sometimes sound a helping u before it in anlaut. Thus *luag* (?) a young eel is pronounced ᴡᴜʟŪåɢ or ᴡᴜʟᴜwåɢ by children.

2. Anlaut *lathach* Làhuc, *ar nís na luiche* er nūs nə ʟɪhə as the mouse does, *lá* ʟȧ.

3. ʟ reduced by vocalic affection retains its guttural timbre but becomes alveolar. *Cluain geal meala* màʟə, *a bhealaigh* ə vàʟə gen. of *bealach*. In such cases the short *a* acquires a tincture of *o* colouring.

4. A slender reduced *l* appears in conjunction with y having the sound of Italian *gl*. *fi ghleo* fī lʸō, *Seàan ó Duibhir a ghleanna* ylaɴə, *fliuch* flyuc.

5. *ll* broad affected may be distinguished from *l* in the same situation. It seems to have retained more of the dental and less of the guttural character. *galla-phuic* ɢàʟə-ғɪk, *ar bhalla-chrith* er vaʟə cri, trembling.

6. *l* has the additional character of nasality in *lámh*. For description it will be enough to say that the nasal passage is open during its production and that it can be best imitated by sounding an *n* before it.

7. *l = n* in *ní fhuil* nīn. The writing *ní bhfuil* which postulates nasal affection can have descended only from *nichonfil*, the sound as far as I am aware is everywhere regular from *nifhuil*.

8. *l* is assimilated to *r*. *dealradh* dauʀ, *siolrach* sīʀuc. In C. M. O. *parrthas báis* 'paralysis' of death comes through *paralios, parlios, parr-*, with a *th = h* developed after *rr*, pàʀəhis. At present its shape corresponds with *parrthas* from *Paradīsus*.

9. *l* takes *d, n,* in assimilation. *muinleach* мūluc, *codladh* koʟə.

10. *ll* broad gave off a svar. vowel in certain contexts. *galla-phoc* P. P. 170, *malla-roisg* ib. 148, *balla-chriolh*.

11. *ll* in auslaut or *l* in position produced slurred diphthongs *mall*, мauʟ, *cill* kil. §§ 4. 6,1.

m

§ 56,1. *m* broad м; slender m. Broad *máthair* мȧhr̩',

mála ᴍᴀ̊ʟə. Auslaut and inlaut from *mm, mb; cam* ᴋᴀᴜᴍ, *am* auᴍ, *iomchur* aᴜᴍᴘʀ´.

2. *m* slender; *mil* mil, *mithid* mihiᴅ (d is often a dental in auslaut). Inlaut and auslaut, *cimil* kimḷ, *ainm* anṃ, *timchioll* hoimᴘʟ´ haimᴘʟ´.

3. *mh* broad in auslaut = v. *cuimhneamh* ᴋInuv, *sámh* såv, *ḍeanamh* dēɴᴜᴠ, dīᴀ́ɴə, *talamh* ᴛᴀ̊ʟᴜᴠ, *mh* slender = v, *nimh* niv.

4. *mh* broad after *l, r* in the auslaut of an accented syllable becomes ᴜ̄ in conjunction with a svar. vowel from those consonants. *ionmhas* iɴᴜ̄s, *ionmhain* iɴᴜ̄n, *seinmhint* sinūnt, *greannmhar* gr´aɴᴜ̄ʀ, *diolmhanach* diʟᴜ̄ɴᴜᴄ, *talmhan* ᴛᴀ̊ʟᴜ̄ɴ gen. of ᴛᴀ̊ʟᴜᴠ, *ullmhughadh* oʟᴜ̄ə from oʟᴜᴠ. So after *ch* in *deachmhadh, deachu,* Cat. 13.

5. *mh* is absorbed after a long vowel sound. *námha* ɴå, *7 tri na laa = lámhaibh,* Cat. 40, so from this a sing. was made, *a la* his hand, Cat. 42, *dream na deas lámha* gen. ʟå T. G. *Mumha* ᴍᴜ̄ Munster, *caomhnach* ᴋᴇɴᴜᴄ, T. G. 10, *ciumhais* kūis, or kūwis; with a slender vowel it becomes y and contracts to ī; *nimhe* nī, *coimheasgar* ᴋīsɢʀ´. So in *Mumha, námha, lámh?* the lengthening was caused by contraction with w. Except in *greannmhar* this termination -*mhar* appears as -ᴠʀ´. *glórmhar, ceolmhar, eudmhar, seunmhar.* In kaʟᴜ̄r ᴍᴜ̄əʀ tī a large rambling old house, there is probably some such word as *Calvaria, cealbhair.* ʙᴀ̊ɴə laᴛ is probably *bana leat* not *b'anamh, snáimhte* is sɴᴀ̊ᴛə. ᴍīᴛə from *mavidheamh, clúdhamhail* ᴋʟᴜ̄əl, *modhmharach* ᴍōʀᴜᴄ, C. M. O.

6. *m* in *chum = n. chun,* Cat. 17, also *chuig* (not in the Dēsi) the prep. pronoun for prep. as *aige* for *ag.*

7. *mh* with broad vowels in accented position gives au. *samha* saᴜ or sauwə § 4,2; unaccented=ᴜ̄. -*amhail* -ᴜ̄l, *greann-mhar* gr´aɴᴜ̄ʀ, with svar. from *nn.*

8. *mh* slender in anlaut=v. *ba mhian liom* ʙᴜ ᴠīᴇɴ lᴜᴍ,=*bh* in the same situation. The anlaut and auslaut sounds of *mh* and *bh* have fallen together just as those of *dh* and *gh.*

9. *mh* slender in inlaut accented makes î and ɛi § 6,1: 8,1; *doimhin* ᴅîn, *treimhse* trɛisə, *deimhin* dɛin.

10. In Thom. Des. sometimes=v where Dēsi contracts. *domhain, dovin,* Cat. 5, *amhairc : deasga* and therefore avʀ′ᴋ, T. G. 9, Dēsi îʀᴋ, *ceanamhail* kaɴəvl̦, T. G. 41. The same is observed in the case of *bh, tabhairt, labhaırt,* ᴛourt ʟourt, or ᴛourt, ʟourt, are sometimes ᴛavr̦t, ʟavr̦t in songs.

11. *mh* broad in anlaut=v or w. *a mhac* ə vâk or ə wâk.

12. *mh* sometimes casts a nasal umlaut back to the beginning of a word, *comhair, comhachta, lámh,* ᴋōr′, ᴋᴜᴜcᴛə, ʟǎv, where ᴋ and ʟ are strongly nasal.

13. *amháin* is əwǎn and *admhuighan* shows a svar. between *d* and *mh,* aᴅəwīm.

14. An -*um* ending is found in slang words, *dubhartum-dártum,* tittle-tattle, *bogadúrum, sancum, mangalum dúd,* a muffler around head and ears.

15. *mm* developed a svar. in words like *camm* before certain anlauts, *cama-chlis.*

16. *m* assimilates *b. camm çambo-, imm* from *imb, dumblas* ᴅuᴍʟ′is. Where h provection enters in the *b* is preserved as ᴘ. *iomchor* auᴍᴘʀ′ from *imb-chor, iompódh* auᴍᴘō from *imb-shoud.* Hence h provection is earlier than this assimilation.

17. *mh* is provected to *f* by h. *liomhtha* līꜰə, *naomhtha* ɴɛꜰə, *neamhdha* nāꜰə on the analogy of *naomhtha.*

18. *mh* or *m* of *mic* is sometimes dropped. *tir ′ic Cláin* C. M. O. 19, *Piarus ic (G)earailt, Seann ′ic Shémuis* S. R. Cf. Welsh and Breton *ab* from *mab, vab.*

r.

§ 57,1. *r* has several pronunciations or variations of timbre. Here only three grades will be distinguished.

r in unaffected anlaut=ʀ. *rath* ʀǎc, *rí* ʀī. *rr* in inlaut is like this, *barra fearra* ʙàrə faꞩə. In auslaut *rr* appears to be unvoiced, *bárr.* From this circumstance would come the tendency to sound an h (expressed by *th*) after *rr* as if it were the result of effort in causing the vocal chords to vibrate again. Single broad *r* in inlaut is hardly to be distinguished from a fairly trilled English *r*. Slender *r* in *-aire* of *faraire*, etc. and in such combinations as *pre*, *bre* is the very reduced sound r´ already described. The *r* of anlaut (where there does not appear to be a distinction of broad and slender) under conditions of affection becomes also r´; *a rí ɔ r´ī, dá rádh å r´å.* In *fuaramar, tángamar* etc. the auslaut *r* is also r´, hånəmr´ (or hånīmr´, Crowley), so *muna mbeidheadh* mr´àc, Crowley. As the same phenomenon is not observable in the case of the 3rd person it is reasonable to conclude that the foregoing labial influenced it. *fuaradar* fŪRɔDR´.

2. In Kilkenny this r´ became zh. *máireach* MåzhUC, *Máire* Måzhə, *bóthairín*, by contraction *bóthrin* Bōsīn, zh to s on account of unvoiced *r* from *rth*.

3. *rr* in auslaut or *r* in position lengthened the vowel under the accent. *fearr*, O. I. *ferr* fāR, *gearr* gåR gāR, *bárr* O. I. *barr*, BåR. Those kept the short vowel on breaking the group; in this case by the addition of a svar. vowel, for the assumption of such hindered the conditions for lengthening. *is míle fearra dhúinn* faRə, T. G. 9, but *malairt is míle fearr* fåR, (Dēsi fāR) ib. 90, *dul na barra. r* in position; *deárna* T. G. 37, *ar bheárnas* ib. 31, *bearna dhearg* a place name bāRNə yaRUG, *deárnacha* T. G. 44, *téarnamh* ib. 77, *athrughadh* åRhŪ, and *aithrighe* århī, where rh constituted position. Those two words are the same, and *aithrige* meant originally change. Vid. *aithirgid bésu* change your customs, Wb. 9a 23. The use of the word in such contexts produced the technical meanings, *change* from sin, and the means of effecting the change, or the Sacrament of Penance. There is also a noun *athrach, ní dhéninn t'athrach* I would not exchange thee, song, but pronounced áhr´uc where *r* sonans hindered position, *cáirde* pl. of *cara, dóirse* pl. of. *dorus, táirne*

a nail, Dēsi ᴛᴀʀɪɴə. So (once) *táirne* for *tarraingthe*, *péirse* Eng. 'perch' C. M. O. 1.

4. That *rr* has a tendeney to develope h is shown by double forms like *orra ortha, siorraidhe* and *siorrthaidhe*. *ortha* is no doubt the etymologically correct form from *-shōs but became confounded with *rr* because that group gave off an h.

5. *r* is a fruitful source of inetathesis. *cearthadh gan lucht*, O. I. *cethir*, Sheehy, *munartla* a sleeve, *brollach* is ʙᴏʀʟᴜᴄ M. song. So from 1, *cuilceach* ᴋɪʟkâc for *cluiceach* from *cluig-theac* bell-house, another name is used in *cluicill Dheaglain* ᴋʟɪkil yīᴀɢʟân .|. *cluig-chill Dēcalāni* in Ardmore in the Dēsi. *trasna = tarsna, bratlin* ʙâʀḷīn, *craosach*, of ruddy face, is not connected with *craos* but is metathesis from *caorsach, Catharlach, Caralegh*, Dunne. Hence 'Carlow' with -ow imitated from Danish 'Wicklow'. In the Dēsi *Catharlach* ᴋâʜʀ´ʟᴜᴄ, *anbhruith* ᴀɴʀ´hə. It is hard to discern whether unvoiced *r* should be written *hr* or *rh*. Hence *liathróid* is often written *liarthóid* and cf. *cearthadh* above.

6. *uathbhásach* is ᴜʀFâsᴜᴄ, *triur* is trʹūʀʀ´ = *triur-fhear*.

7. *r* is sometimes dropped in *creidim* kedim, so *iomarcradh* is *iomarcadh* always. Also *chondairc* is ᴄɴɪk.

s

§ 58,1. *s* is never voiced. Broad = s, slender = s. It occurs in anlaut and protected in inlaut, otherwise it becomes h. *solus* sᴏʟis, *searb* sᴀʀᴜᴠ, *is* is; probably this is but the relative form *as*. In enclitic particles it changes its timbre according to the foregoing vowel, *annsin* ɴ´sᴜɴ (here accented), *an fear sin* sɴ´, but *an fear sin annsin* sin, *mise tusa* ᴛᴜsə, an old change. Hence the writings *seo so, san sin*. *sin* a mere enclitic, not demonstrative, = sɴ´; *thád siad go dona, thád sin* sɴ´, they are in a bad way so they are.

2. *ó shoin* is ō ᴄɪn, *cá shoin* ᴋâ or ᴋà ᴄɪn. § 10,1.

3. *s* of the article kept after preps. *in, la*, etc. has spread to other preps. in the pl. *fésna, trísna, dosna, desna (gesna)*. A wrong division of the prep. *in* from the article

gives *ans gach*, T. G. 17. So *s* preserved after *ar* 'quoth' was used without the article, *ar san fear arsa sé* says the man says he, *arsise* and *arsa sī*.

4. An -*is* = is ending, mostly feminine, appears in a few cases. *ótis* a heavy ungainly woman, *brocis* a stout little boy, *broc* a badger, *tulcis* a rude woman, *tulc* a push, O'R. *alpis* a gluttonous woman, *alpim*, I eat ravenously, *balcis* a stout mis-shapen woman, *balc*, thick, strong, *strupis* an untidy woman. *lítis* a lily, *líth* D. R. 178, *buatis* is only Eng. 'boots'.

5. *sg* + *l*, *r* suffers palatalization with slender vowels *sgeul*, *sgléip*, *sgread*, sgr-.

6. *s* of *súd* and *sin* = h when those words are proclitic in sentence anlaut. *súd é* hiDē, *sin é 'n choint* hinē N' coint.

7. *sh* before *y* = c in a few words. *a Shaáin* ə căn, or ə hyăn, *luighe-sheol* childbed cōL, *do shiubhal* ə cūL, or hyūL, otherwise = h, *shlighe* hlī.

8. *s* voices *t* in inlaut, *steall* sDauL. *stealladh* sDáLə, in auslaut unvoiced, *last* LásT.

9. *s* in anlaut, *sd* in inlaut and *sc?* in auslaut takes the place of an affricate in Eng. loanwords. *Seóirse* George, *lóisdín* lodging, *cisdin* kitchen, *damáisde* damage, *pitrisc* partridge, *carráiste* carriage, *páiste* page, *Sémus* James, *Seaán* John, *cóisde* coach. *Seóirse* shows irregularity of treatment. *Risteird* risterD, in the West RUKR'D.

h.

§ 59,1. *h* is pronounced wherever met with except it be but the aspiration sign. *a h-aon* ə hEN, *ní h-eadh* nī ha.

2. *h* is sounded for *f* in fut. and cond. active after a vowel or a consonant that cannot be provected, *ni sgarfad leat* ni sGáRhəD laT.

3. *h* interchanges with *f*, *b*, § 47,5.

4. *h* unvoices *l*, *n*, *r*. *shleas*, hlas, *do shnámh* hNăv, *aithrighe* ăhrī, *eatortha* aTṛhə.

Curriculum Vitae.

In Hibernia Idibus Septembris 1863 in regione quæ vulgo Duthaig Poerach de Desibus Muman nuncupatur Ædibus manorialibus de Mount Bolton patre Perso de Henebre matre Eblina ni Chassin natus sum. Postquam domi elementis institutus sum parvam scholam primum Carrigæ Siuiræ deinde Cluain móræ tum Portcladaci adii. Hisce in disciplinis puerilibus decem annis consumptis Portlárgiam (Waterford) me contuli ut litteris humanioribus per biennium incumberem. Tum alumnus Collegii Sti. Joannis eadem in urbe ascriptus sum easdemque ad litteras per unum annum me applicui, et examine facto electus sum ut sumptibus publicis in Collegio S. Patricii apud Mag-Nuadat studia prosequerer. Hic septem persolvi annos quorum unum ad humaniora studia, ad philosophiam tum physicam tum metaphysicam duos, ad theologiam, Sacram Scripturam, grammaticas Hebraicas, studia affinia, quatuor me adhibebam. Nec tamen hisce temporibus neglexi litteras aut Anglicas aut Gadelicas quas utpote bilinguis ab ineuenti aetate usurpavi. Mense Junii 1892 sacris initiatus ut munere sacerdotali Mancuniæ (Manchester) fungerer in Angliam transfretavi. Hic Joanne Strachan in Universitate Victoriana professore, viro doctissimo atque humanissimo familiariter usus sum eundemque per quadriennium de Sanscritis, Celticis litteris audii. Mense Aprili 1895 professor litterarum Celticarum Universitatis Catholicæ apud Washington in America designatus, præside Senatus auctore ad Germaniam transii ubi per quatuor semestria, quorum duo Friburgi, Gryphiæ duo, persoluta, studiis operam dabam.

Friburgi hi viri clarissimi me docuerunt:

Holtzmann, Kalbfleisch, Kluge, Levy, Puchstein, Schröer, Thumb, Thurneysen, Weissenfels.

Item Gryphiæ:

Heller, Konrath, Siebs, Zimmer.

www.ingramcontent.com/pod-product-compliance
Lightning Source LLC
Chambersburg PA
CBHW021426090426
42742CB00009B/1270

* 9 7 8 3 7 4 4 7 3 4 5 2 3 *